Dealing with
RISK

*The planning, management and
acceptability of technological risk*

Dealing with
RISK

The planning, management and acceptability of technological risk

editor

Richard F. Griffiths

A HALSTED PRESS BOOK

JOHN WILEY & SONS
New York

Published in the USA by
Halsted Press,
a division of
John Wiley & Sons, Inc., New York.

ISBN 0 470-27136-1

Typeset in Great Britain by Express Litho Service (Oxford)
Printed and bound in Great Britain by
Biddles Ltd, Guildford and King's Lynn

Contents

List of contributors

C. W. Brough
Head of Planning Services, Halton Borough Council,
Kingsway, Widnes WA8 7QF

A. V. Cohen
Planning Unit, Health & Safety Executive, Baynards
House, 1 Chepstow Place, London W2 4TF

Professor Stephen Cotgrove
School of Humanities and Social Sciences, University of
Bath, Claverton Down, Bath BA2 7AY

Richard F. Griffiths D.Sc.
Assistant Director, UMIST/University of Manchester
Pollution Research Unit, University of Manchester
Institute of Science and Technology, Manchester
M60 1QD

T. A. Kletz
Division Safety Adviser, Imperial Chemical Industries
Petrochemical Division, Wilton, Middlesbrough,
Cleveland TS6 BJE and Industrial Professor, University
of Technology, Loughborough

J. McLoughlin LL.M.
of Gray's Inn, Barrister-at-law, Faculty of Law,
University of Manchester, Manchester M13 9PL

B. J. Payne
Cheshire County Council, County Planning Department,
Commerce House, Hunter Street, Chester CH1 1SN

Sir Frederick Warner F.Eng. FRS
Cremer and Warner, 140 Buckingham Palace Road,
London SW1W 9SQ

Foreword: The foundations of risk assessment

Sir Frederick Warner

Introduction

The major theme of this book is the acceptability of risk, as distinct from the assessment of risk itself. However, these two aspects are components of the overall field of risk management and I will concentrate my remarks on the subject of risk assessment and the foundations on which we base our quantified estimates. This is an area in which I find, personally, very considerable philosophical difficulties, some of which have their origins in problems of terminology and definition. This kind of problem is well illustrated by the difficulties encountered in translation during international discussions. Limitations to one's understanding arise because of failure to comprehend the exact meaning of words, particularly in the case of those that *look* to be the same. Consider, for example, the book by Jacques Monod *Le hasard et la nécessité*. It is tempting to assume that *le hasard* corresponds to the English *hazard,* but it is actually *chance*. We find similar difficulties in the field of risk where many different meanings of the same terms are apparent.

The importance of agreeing on definitions and terms leads to consideration of the use of codes of practice and standards. On this subject I have long held the opinion that in risk assessment and in safety generally it is a very good thing for people to begin from what is known. The knowledge incorporated in codes of practice and standards represents in many respects the best information available at the time of drafting, and care is taken to ensure that the information is in a form which is useful to people operating in the field. Too often these standards are not well enough known. It will be remembered, for example, that during the Flixborough Inquiry it became clear that nobody on the site at Flixborough was familiar with the British Standard then in existence for piping in petrochemical installations which specifically prohibited the type of by-pass pipe that was installed between reactors 4 and 6. In fact the standard was infringed in many respects. One of the criticisms that emerged was that there was no engineering organisation independent of production-line management responsible for assessing the overall system and ensuring that proper controls were exercised.

The importance of such standards is clearly demonstrated by the above example; in the field of risk I would commend the fundamental definitions which are set out in the British Standard *Glossary of terms used in quality assurance* (BS 4778:1979), especially Section 13 on safety and Section 14 on probability. This document is one of many signs of the current emergence of a probabilistic approach to the treatment of reliability in engineering which is replacing the older factor-of-safety practices.

An important distinction is brought out in BS 4778 (Section 13) where it deals with risk assessment as incorporating *risk quantification* and *risk evaluation*: 'This separation can be beneficial in avoiding confusion between the objective process of risk quantification and the essentially subjective interpretation of the significance of estimated risks.' It then goes on, with special significance to this book's subject:

Risk quantification cannot measure risk acceptability, which is the concern of the relevant decision makers who have to judge the benefits, alternative uses of resources and other factors unrelated to the process of risk quantification. Moreover, the uncertainties in the quantification do not become entangled quantitatively in the process of judgement. They are simply determinants of the potential range (or perhaps distribution) or variation of the value of the quantified risk. This range can be assessed for acceptability.

I believe that this document provides a useful point of reference in establishing the terminology that we should use in this field. Additionally, it has the advantage of being to a considerable extent a pioneer document, in advance of whatever is eventually agreed internationally by the International Electrotechnical Commission and the International Standards Organisation. Furthermore, it represents the views of manufacturers and operators as well as relevant government departments, who have conferred through the usual process of consensus which precedes the drafting of British Standards. I think, too, that there is no real possibility of our achieving any real reversal of the decline of British industry except by adopting the standards followed by organisations that implement these procedures, who understand the importance of quality assurance and reliability in production.

Public opinion and risk perception

One of the topics dealt with in this book is the perception of risk. I find this one of the most difficult aspects of risk management and it is clear that people involved in examining public perceptions approach risk from a point of view quite different from that of the engineer. I recall that when the Windscale Inquiry Report was

published, Professor Cotgrove wrote to *The Times* saying that the participants were really not discussing the
issues on the same basis and that the discussion should
really be about values not facts. Again, we encounter
criticisms of engineers for wanting to deal with what we
might call hard-edged numbers. I am not implying that
the findings reported by sociologists are particularly soft-
edged, but I do not think one can proceed even with the
assessment of values and the discussion of acceptability
if the argument has to be decided purely on the basis of
opinion polls. Given the variations of opinion that occur
even over a period of months, one could not justify
basing long-term decisions on such changeable information.

Whilst on this subject I think it is highly instructive to
examine the different ways in which governments react
to public opinion, because one can see widely differing
approaches in different countries. Taking nuclear power
as a topical issue in this respect, the 1977 Windscale
Inquiry was really a first attempt to engage in a very
wide-ranging public discussion with as much assistance
as possible given to all the objectors. The Inquiry sat for
100 days hearing submissions from many witnesses; the
written evidence produced a 4½ metre stack of paper
and every day there was 45 mm of new reading for everyone directly involved. I would estimate the final cost to
be about £2 million to £3 million, but I sometimes
wonder whether the inquiry was in fact as good as it
might have been. Now, while the Windscale Inquiry was
going on in the summer of 1977 there was a demonstra-
tion by protesters in France against the siting of the fast
breeder reactor at Creys-Malville. Here we see one indi-
cation of the difference in approach between the British
and the French in dealing with public opinion; the
French turned out 10 000 riot police, at a probable cost
of £200 000, one man was killed, and very little more
was heard. That is just one example. Another is the way
in which they organise their process of consultation.

The French have official bodies, like Electricité de France, supervised under the Commissariat à l'Energie Atomique; their process of consultation consists of meetings with people living within 5 km of the site and, since the decision is made centrally, the public at large do not really receive a lot of information. It is in some ways frightening to contemplate the way in which the Framatome programme for commissioning 1 gigawatt every 3 months up to 1984 has been achieved so far.

Examination of public perceptions of risk yields all manner of surprises. The Windscale Inquiry revealed an odd effect in that the perception of risk appeared to increase with distance. We had lots of people from Keswick who were worried about the genetic effects of radiation, which should not really be a topic of major concern because the risk is very small, whilst people further afield in Oxford were paying a lot of attention to the source terms in the atmospheric dispersion model for releases of krypton 85. This conflicts with what one would expect and with one's experience in that the perception of risk usually falls with distance in a way analogous to the laws of attenuation for noise or explosion damage. For example, 18 months after the Flixborough explosion, in which 28 were killed, there was a refinery explosion at Beek in Holland, killing 14, but very little attention was paid to that event in the UK, nor was there much concern apparent just across the border in Germany. The same applies to the Gujarat dam failure in 1979, in which 15 000 were killed. With this number of fatalities one really is dealing with a major hazard and yet there was very little attention paid in the UK beyond a modest coverage in the news media. Apparently, being 3000 miles away from the site of the disaster reduces the public perception of the significance of this event to a very low level.

Uncertainties in risk assessment

Turning again to this question of criticisms levelled at engineers and the confidence placed on their numerical estimates of risk, I think we have to confess that these numbers are far from being hard-edged. These estimates are derived from all sorts of sources and are used in many different ways, so they certainly do have different degrees of precision. A great deal of thought has been expended in estimating the degree of precision or the range of uncertainty attached to risk estimates. The problem is well recognised and is made explicit in the use of the various kinds of information on which we draw, as I will illustrate with three examples:

(*a*) Looking first at mortality statistics I will refer to the individual risk survey prepared by Grist (1978). There are two sources of uncertainty involved in this sort of statistic. One is the systematic error that arises because of social and environmental effects on the incidence of death. A specific example would be in the statistics on smoking and its effects on health. Fox & Adelstein (1978) have examined the correlations that exist and how these are affected by social and economic factors. The implication is that mortality tables alone are not a great deal of use as indicators of risk but need to be interpreted in the light of these other factors. The second source of uncertainty arises where the number of deaths in a particular category is small, in which case random statistical fluctuations are very important. Grist points out this problem and discusses it in terms of the confidence limits of the expected value of the Poisson Distribution. Grist's paper is very useful because it classifies the data in various ways, according to age, cause of death and so forth. It is also the first example I have seen where the WHO International Classification of Diseases has been used to group the data. I would recommend this report as being well worth studying.

(*b*) The second class of information we rely upon, especially in the risk assessment of major hazards, is that contained in data banks on reliability and failure rates. The major bank of this kind is the Systems Reliability Service (SRS) run by the United Kingdom Atomic Energy Authority (UKAEA) at Culcheth. There are other individual banks held by organisations like ICI, the Central Electricity Generating Board, the Ministry of Defence and the large oil companies. Access to most of these is available through the SRS system. Failure-rate data have to be classified in various ways, not just on the basis of the type of component, because some failures are only partial and are capable of reversal (in this connection I will refer you again to BS 4778 where hazards are divided into four groups, catastrophic, critical, controlled or marginal, and negligible). These data banks are continually being added to and provide very important information drawn from experience and processed so as to include estimates of confidence limits and ranges of uncertainty. These uncertainties enter the full analysis of hazard. Taking an example from the Canvey Island Study (Health and Safety Executive, 1978), the effect of releasing 1000 tonnes of liquefied anhydrous ammonia as the result of a catastrophic failure of the pressure vessel is represented as the worst case. Taking category D weather conditions (the most prevalent in the UK) with a windspeed of 3 metres per second, the distance within which fatal concentrations would be experienced is calculated to be in the range 5 to 8 km. That is the consequence of such a release resulting from the failure of the pressure vessel, but the likely vessel failure rate computed from historical data is 13 per million per year, with 95% confidence limits of 60 and 2.8 as the upper and lower estimates about 13. That is quite a range, and when combined with all the other sources of uncertainty the result is very far from what I would call a hard-edged figure.

(*c*) The third set of information to which I refer is the evidence on risk to health available from epidemiological studies. This is a difficult area for purposes of risk assessment because there is always the uncertainty about establishing the causal link when dealing with correlations. For example, there are arguments going on to this day about the effect of smoking on health. It may even be that the case for giving up smoking is not always so straightforward as one might suppose. I have seen a paper that argued from statistical evidence that the risk of death amongst medical doctors who had given up smoking was increased because they were then more susceptible to stress diseases, as shown by the increased incidence of suicide amongst those who had given up smoking. Of course smoking is a voluntary act and therefore the risk is different in nature from one that is imposed on you, but the health risks of smoking are often emphasised in comparisons of various forms of risk. We are told that the individual risk of death is 1 in 400 per year for a person smoking 10 cigarettes a day (Royal Commission on Environmental Pollution, 6th Report, 1976) or, to put it in other equivalent ways, that the life of an individual is diminished by the time that he spends smoking or that smoking just 1½ cigarettes represents a 1 in a million risk of death per year. However, I think that any engineer confronted with numbers like those would be immediately suspicious because we do not believe that these figures are really so clear-cut.

Dose—risk relationships and occupational risk

Implicit in the foregoing statements about the risk of smoking is an assumption that for every hazardous material one can apply a dose—commitment model. Now the dose—commitment approach is taken over bodily from the field of radiological protection. The way in which this is applied can be illustrated by reference to the Three Mile Island accident. It is estimated that the population dose received was about 4000 man-rems. Given a dose—risk relationship of approximately 1 excess cancer per 10 000 man-rems, the outcome is a prediction of about one excess cancer over the next 30 years amongst the exposed population. (To put that in perspective a dose of 4000 rems is typical of what one person would receive from the radiological treatment of a small skin tumour). This illustrates the way in which we apply the linear dose—response relationship for effects of radiation on biological systems, but any attempt to apply the same concept in the case of chemical effects should be regarded with extreme caution. This problem is discussed by Pochin (1978), from whom I quote:

The use of a linear model for a dose—effect relationship in setting radiation standards seems to have a rather limited reference to chemical carcinogens because its use as an empirical model would not depend on any analogy with radiation but on its reasonableness in the absence of better information, and because the increasing information about the validity of limitations of linear extrapolations for radiation is very specific to the facts and theories on the mechanism of radiation effects on cells. It may be noted, however, that for many chemical pollutants, the metabolic products which are carcinogenic may have zero tissue concentrations under 'normal' circumstances. This would, if so, differ from radiation increments to natural radiation exposure and imply that chemical repair mechanisms might be more likely to impose a 'threshold' below which no effects occur. The major difference between the two situations (radiation and chemical carcinogenesis) seems to lie in the contrast between the physically predictable nature of the radiation dose to particular tissues or cell structures, given adequate knowledge of the normal metabolism of the ionic or molecular form in which radionuclides are taken into the body, and the less certain knowledge of the way in which many chemical pollutants may become distributed and metabolised within the body and at the cellular level. It is easy to imagine ways in which the difference could drastically affect the validity of extrapolating from high to low dose effects — for example, either by small amounts of metabolically unusual chemicals saturating enzyme or immune systems of small capacity, so that high doses were no more effective than low doses; or by low doses of such chemicals becoming metabolically degraded so that a threshold occurred below which no effects were produced.

That of course raises a question about the meaning of one of the control schemes used by the Health and Safety Executive, namely the threshold limit values (TLVs) derived by the American Conference of Hygienists. TLVs are intended to be applied where there is continuous occupational exposure for eight hours a day with a five-day working week. But what are we to understand by the use of the word threshold in this context? Does it mean that there is a level of exposure below which there are no effects, or does it mean that the effect is so low that it disappears altogether? For many chemicals the TLVs will often be indistinguishable from the noise, the range of uncertainty. However, the merit of TLVs is shown pragmatically by the actual incidence of damage and by the trends apparent in data on industrial safety. None the less, there are still great problems to be tackled in establishing the comparisons that need to be applied so as to give proper allowance for differences in the kind of work undertaken in different occupations, the age at which exposure takes place and all the other factors thought to be significant.

Another very difficult problem concerns whether or not some fates are worse than death. Most statements about risk are in terms of immediate or delayed deaths, but there is certainly the possibility of total or partial disablement which clearly detracts seriously from the quality of life. As yet we have no real means of incorporating this in our data on detriment. Sir Edward Pochin (ICRP 27, 1977) has examined the incidence of deaths and injuries at work using worldwide data from various occupational health records. He found that for the United States data the accidental death rate, D, was approximately related to the rate of occurrence of disabling accidents, A, by the expression

$$D = kA^2$$

where k is a constant.

This means that the death rate increases more rapidly than the disabling accidents rate, or, alternatively, in the more hazardous occupations fatalities constitute an increasing proportion of the total harm caused. He also found that the mean loss of life per occupational fatality is about 30 years and that this figure increases at about the same rate as the mortality rate falls in a given industry. If it were to be assumed that the period of time lost through being dead was regarded as worse than an equal loss of time through being disabled, then accidental deaths would constitute the major contribution to harm in conditions where the time lost per death was more than 30 years. This is very difficult ground indeed because in some cases death is regarded as preferable to certain kinds of injury.

The Japanese data examined by Pochin show that *total* disability occurs at about 5% of the death rate from industrial causes, but permanent *partial* disability occurs at about twice the death rate. The general conclusion is that the occupational fatality rate measures the major contribution to harm for occupations with fatality rates greater than about 200 per million per year, except where there is a high incidence of industrial diseases.

An approach related to that adopted in ICRP 27 is explored in the paper by Reissland & Harries (1979) who have presented comparisons of the average loss of life-expectancy for individuals engaged in various industries. Their comparisons are expressed as the average number of days of life-expectancy lost as a function of age at the beginning of exposure to the risk, for one year's exposure and for exposure over the remainder of the working life. To illustrate the range of values obtained in this treatment, the figure for a worker who engages in deep-sea fishing throughout his working life from the age of 20 is about 1400 days, whereas the corresponding figure for a radiation worker exposed to 0.5 rem per year is about 7 days, a loss which is hardly significant against the background of normal life-expectancy.

Conclusions

I hope that the examples I have given are sufficient to demonstrate that the figures expressing the results of an engineering assessment of risk are really very far from firm. I hope, too, that the views I have expressed will have some impact on those who approach the subject from a different point of view and who criticise the engineers in their attempts to quantify risk. I am sure we shall face more and more surprises as we go on trying to establish these foundations for the assessment of risk. When thinking about whether we have a reliable basis for this endeavour I come back to what is almost a theological position and one which was stated by a man who was both a theologian and a founder of the science of fluid mechanics, Blaise Pascal. I translate loosely from his *Pensées*:

All our foundations crack. We are on fire to find this firm plat-form and a base on which we can raise a structure which rises up to the infinite — our foundations crack and the prospect that opens to us is that of the abyss.

This book deals with a subject area in which engineers do not feel especially comfortable because they are asked to provide figures on estimates of risk whereas the judgements that have to be arrived at are made by other people. One needs to be wary of making predictions; we know from historical data that about once every ten years we have a railway accident in the UK in which 100 or more people are killed, but on a certain occasion last year I ventured to say that the one thing you can gua-rantee on British Rail is that two trains will not collide head-on. Two weeks later that very thing happened at Paisley. In the aftermath of such an event there is the considerable danger that too many additional safety measures are imposed on the railways, forcing more traffic on to the roads where conditions are really a great deal worse and the risks higher.

The lack of correspondence between people's perceptions of risk and the corresponding engineering estimates is a problem which I think is unlikely to be resolved. Given this situation I believe our response should be to try always to put information before the public. I am a great believer in the freedom of information. Of course, industry is somewhat constrained in this respect by considerations of commerical confidentiality, but the more that can be done to inform the public the better. If we do this and admit that in the quantification of risk our calculations are based on foundations which are fairly unstable, so that the figures are subject to quite a degree of uncertainty, then that I think is all that we could or should do.

References

British Standards Institution (1979). *Glossary of terms used in quality assurance (including reliability and maintainability terms).* BS 4778.

Fox, A. J. & Adelstein, A. M. (1978). Occupational mortality. *Journal of Epidemiology and Community Health,* **32**, (2).

Grist, D. R. (1978). *Individual risk — a computation of recent British data.* UKAEA report SRD R125. Her Majesty's Stationery Office, London.

Health and Safety Executive (1978). *Canvey — an investigation of potential hazards from operations in the Canvey Island/ Thurrock area.* Her Majesty's Stationery Office, London.

Pochin, E. E. (1977). *Problems involved in developing an index of harm.* ICRP 27, Pergamon Press, Oxford.

Pochin, E. E. (1978). Assumption of linearity in dose—effect relationships. *Environmental Health Perspectives,* **22**, pp. 103—5.

Reissland, J. & Harries V. (1979). A scale for measuring risks. *New Scientist,* **72**, pp. 809—11.

Royal Commission on Environmental Pollution 6th Report (1976). *Nuclear power and the environment.* Her Majesty's Stationery Office, London.

1 Introduction: The nature of risk assessment

Richard F. Griffiths

There must be only a very few real problem areas that can be adequately treated within the confines of a single discipline, and the subject of technological risk (as distinct from financial risk) is certainly not one of them. Whenever members of multidisciplinary groups meet to discuss their contributions to a field in which no single discipline can truly claim to be the leader one expects vigorous debate. In the case of technological risk there are two additional factors that have given extra impetus to the developments that are now emerging: widespread public concern over the risk implications of our commitment to large-scale industrial technology, and the response of central government arising from its awareness of the public concern. This whole field is still very much in a state of flux but a pattern is beginning to emerge that characterises the scope of the subject. As elaborated in Dr Cohen's paper, three major components seem to exist:

(a) *The quantitative assessment of risk*, consisting of an engineering/scientific exercise to identify potentially hazardous events associated with a given project and to estimate the risk in terms of the likelihood of occurrence and the severity of the consequences.

(*b*) *The decision* on whether or not the risks are such that the project can be pursued with or without additional arrangements to mitigate or regulate the risk.

(*c*) *The legitimisation of the decision,* i.e. is the decision acceptable to society?

The papers presented here are mainly concerned with the last two items, in that we did not deal with details of the techniques of quantitative risk assessment. However, the distinction between technical and social aspects is not so readily maintained in this subject because whilst much of the technical procedure of risk assessment relies on established science or engineering knowledge, there are significant elements that depend on the exercise of what are known as 'engineering judgements', i.e. intelligent and informed guesses.

Those responsible for undertaking specific risk-assessment projects (as opposed to the formulation of general methodological frameworks) would prefer to have available unassailable methods for all the calculations they are required to make, but unfortunately this is not the case. None the less, when such projects are commissioned, individuals are charged with these responsibilities and there is considerable pressure on them to provide 'best estimates' where established techniques do not exist. There are legitimate reasons why such judgements need to be invoked:

(*a*) The knowledge of a particular phenomenon which figures in the overall assessment may not yet have been developed beyond the fundamental research stage, in which case reliable predictions of sufficient accuracy may not be possible, e.g. the atmospheric dispersion of toxic or explosive mixtures that are denser than air.

(*b*) The knowledge required is obtainable in principle but it may not be ethically acceptable to undertake the particular investigations that would be required to yield accurate information, e.g. effects on humans of certain toxic agents.

(*c*) Uncertainties in the validity of extrapolating from laboratory to industrial-scale systems may have to be accepted because of the economic impracticability of undertaking a series of full-scale tests, e.g. large pressure-vessel failure mechanisms.

This list could be extended considerably, but the general point is that the firm basis of consistent agreement between hypothesis and experimental verification that one seeks in scientific and engineering activities may not be available in certain topics that must be included in a particular risk-assessment project. The idea that one has to deal with uncertainties in quantitative assessments is of course not new to the professional technologist; his training will have included material on uncertainties and sources of error and it will be routine practice for him to make these explicit in his work.

The technologist's familiarity with the concept of uncertainty is perhaps not as widely understood as it should be. Clearly there are different categories to be considered, ranging from simple inaccuracies arising from the limitations of measurement techniques to the grosser inadequacies of knowledge of the kind listed above. This is very much at odds with the persistent popular view of science and engineering as being always involved with *exact* numerical values, and it can only be a good thing that this assumption has been revealed as erroneous through public involvement in the debate on risk. The result of such exposure is that the relationships between technical experts and other groups, especially the public, are undergoing substantial change and a more discriminating view is emerging. The difficulties that arise when experts disagree on technical matters are familiar enough in the closed context of a particular specialist group, but the fact that technical experts display such a lack of consensus in some fields has perhaps come as something of a surprise to the outsider. What may also have come as a surprise, but to the technical experts this time, is their lack of success in resolving some of these problems. Harvey Brooks (1972) distinguishes between *science* and *trans-science* in this connection, the latter term being used to characterise problems which can be expressed in the methodology and language of science but which are not amenable to

resolution within current paradigms. Recognition of this
distinction and its implications can be a painful matter,
but the development is a healthy one in that it leads to a
better appreciation of the nature of science and technology and the way in which they serve society. Sir
Frederick Warner addresses this problem in his foreword;
his frank acknowledgement of the inherent limitations
involved will be seen as a valuable basis for improved
understanding of the relationships and roles of the
various public and professional interest groups.

A criticism that is often levelled against expert
opinion in the debate on risk is that it is irremediably
biassed in favour of the proponents of high technology
by virtue of the institutional arrangements by which
such experts are funded. Many of the opponents of the
Windscale development expressed the view that funds
comparable to those available to the applicants should
be put at their disposal. One wonders to what extent it
could prove generally feasible to devote national resources to such an exercise, but to the extent that facilities funded by public money are involved the point is
a valid one. Practical recognition of this point was apparent during the Windscale Inquiry when TIRION, a
computer program developed by the Safety and Reliability Directorate of the UKAEA and used in the risk
assessment of the Windscale installation, was made available to the Political Ecology Research Group, who were
provided with the program complete.

Such examples serve to extend understanding of
technical issues and methods involved and can be
applauded as such, but it is probably erroneous to
suppose that a campaign designed to improve public
appreciation of technical aspects of risk would do much
to relieve the tension that is currently apparent. This is
not to say that efforts to extend technical literacy
should be restricted, but rather that their purpose
should not be misunderstood. If, as Professor Cotgrove
suggests, the bitterness of some of the exchanges to

which he refers is due to a fundamental incompatibility
between rival systems of values then no amount of
public information can be expected to resolve this conflict. Anyone who had assumed that the problem of risk
acceptability could be solved by such methods will find
this idea discomforting, but there is no reason why this
issue should be exempt from the attentions of strongly
held and conflicting views.

*The necessity for public representation on risk
questions*

The task undertaken by those involved in quantitative
risk assessment is to analyse the incidence and severity
of actions which lead to damage. This includes events
that have occurred, but the analysis of the *potential* for
damage is included and has come to the forefront in
recent years as a topic having widespread implications
beyond the technical.

Immediate injury is only one aspect of the risks
involved in activities based on large-scale technology,
and recognition of this fact has broadened the issue of
safety to include, for example, long-term effects of
routine or accidental exposure to chemicals and the
question of how the burden of risk is distributed over
the population at large. This development has called for
the application of probabilistic methods to the task of
risk assessment and these techniques are making increasing inroads on the more traditional factor-of-safety
approach.

To illustrate the difference between these two
approaches and to show how the probabilistic approach
complicates the issue of acceptability I offer the following scenario: Suppose that a rural community wishes to
attract a greater number of people to enjoy the pleasures
of walks in their region. There will be a financial benefit
for the locals. It is recognised that the walks will be

more attractive if simple bridges are erected over the
many muddy crossings of the streams in the area. Given
the expectation that the majority use would be by
adults, the solution might be to lay planks just strong
enough to withstand, say, twice the weight of an average
adult, and to erect signs advising that only one person at
a time should be on the plank. However, on rare occa-
sions someone might miss the notice and try to ride a
horse across the plank, which would then break. Thus
there would be a distribution of loads to which the
plank might be subjected. The probability of such a high
load may be quite small, but it exists. The strengths of
individual planks will vary too, even with careful selec-
tion. A few will be very much stronger than the majority
and some will be relatively weak. So there would be a
distribution of yield loadings associated with these
planks. It would be possible, of course, to build very
strong wooden bridges, capable of supporting the weight
of a farm tractor, but the cost of labour and materials
would be much higher and could price the whole project
out of the market. Given some information about how
strong the planks are and what pedestrian traffic may be
expected we could give a quantitative estimate of how
often and how far the high end of the load distribution
will encroach on the low end of the plank-strength distribution, arriving at an expected value of how often
people may be dumped in a stream. Clearly, a decision
on how strong to make the bridges is called for, and it
involves many people with different personal interests
who will emphasise different aspects. It would not be
surprising to find that they were unable to agree on how
to proceed. It would equally be very surprising if they
were happy to have the decision made for them by the
local timber merchant who owns 100 planks and who
has been offered £10 for each footbridge that he erects.
There is no one person or group obviously better quali-
fied than the rest to take this decision and democratic
principles demand public representation.

This illustration suggests that it would be unreasonable to place decisions on risk in the hands of one interest group alone. To determine the extent to which one group actually influences an area of concern may not be possible, but it may be just as important to consider the extent to which that group *is thought* to hold such influence. In the field of risk the body of literature recording the efforts of those involved in quantitative assessment is truly enormous. Public attention has been focussed on this aspect in recent years by a number of events, reports and inquiries in which the role of the technical expert has been of great influence. Is it possible that these frequent appearances have promulgated a popular myth concerning the use and reliability of quantitative analyses of risk? I think it probably is the case and would like here to examine the nature of this myth, the way it has developed and how the balance may be restored.

Milestones in the UK debate on risk

Windscale, 1957

The accident that occurred at Windscale in 1957 has a special significance in the UK. Many people recall the ban on consumption of cows' milk that was imposed in the area for a few weeks to safeguard the population from the effects of ground contamination by iodine—131. The quantity of material released in the accident was relatively small (about 33 000 Curies of mixed fission products), but the impact of the event in the public mind was very great indeed. The reactor itself was a once-through air-cooled device designed for the production of plutonium for military use. This kind of reactor is different from those used for the civil production of electricity, especially in its cooling arrangements; a closed coolant circuit is used in the civil design and fission products cannot escape directly to the atmosphere.

For a variety of technical reasons of this kind the Windscale accident is not representative of what could happen in a civil electricity-generating reactor, but this accident is strongly associated in the public mind with nuclear power.

The consequences of the event were minor in terms of actual harm to the population, but the repercussions were great. A Committee of Inquiry was set up (Atomic Energy Office, 1957) and, in response to the concerns expressed at the time, a whole new branch of the Atomic Energy Authority was established to be responsible for reactor safety matters (this was the Authority Health and Safety Branch, later to become the Safety and Reliability Directorate of the UKAEA).

Flixborough, 1974
In this accident, about 40 tonnes of cyclohexane leaked from a failed by-pass pipe on a chemical plant run by Nypro (UK) Ltd. The subsequent explosion destroyed the works and 28 people were killed. This time it was the chemical industry that was the focus of attention, and the consequences have been tremendous. A Court of Inquiry was set up, and its findings (Factory Inspectorate, 1975) were strongly critical of the operators of the plant for failing to adhere to codes of practice and standards laid down for pipework installations. The Advisory Committee on Major Hazards was established shortly after the accident to consider the safety problems associated with large-scale industrial premises conducting potentially hazardous operations, other than nuclear installations. The first and second reports of the ACMH (Health and Safety Commission, 1976, 1979) take a generic approach to the problem and have been highly influential in establishing new research and regulatory initiatives. The Health and Safety Commission have proposed draft regulations (1978) governing the operations of installations where hazardous materials are handled in large quantities.

The influence of these two accidents, and the response of government in establishing new institutions and regulations in their wake, has been paralleled by three major reports involving estimates of risk in advance of the events themselves. Again it is the nuclear and chemical industries that are the focus of attention.

The Rasmussen Report, 1975

In 1975 the United States Nuclear Regulatory Commission published the now famous Rasmussen Report on accident risks in US commerical nuclear power plants. This was a major study involving the expenditure of about $4 million and the deployment of 70 man-years of effort over 3 years. The importance of the Rasmussen Report as a document publishing *a priori* estimates of risk can hardly be overstated. There has been widespread international response, including criticism by various groups who questioned the validity and accuracy of the methods used. There has been a continuing debate in the UK over the possible British adoption of one of the light water reactor systems analysed in the Rasmussen study. Extra impetus has been injected by the occurrence early in 1979 of the accident to the nuclear reactor at Three Mile Island in the United States. Although there were no casualties, international public awareness of the disruption caused by this event has been considerable. The findings of the Kemeny Commission of Inquiry (President's Commission, 1979) were highly critical of the regulatory arrangements in the US and this has added to public concern over nuclear power risks. Operator error is thought to have contributed significantly to the TMI accident and this has led to renewed concern over the adequacy of analyses of risk that concentrate on the system itself. Clearly, it is of great importance that a significant element of risk may arise from a source that cannot readily be included in risk estimates.

The Windscale Inquiry, 1977
In 1977 the Windscale Inquiry was rarely out of the public eye. It was set up under the procedures that apply to specific planning applications. In this case the applicant, British Nuclear Fuels Ltd, wished to build a nuclear fuel reprocessing plant at Windscale in Cumbria. Perhaps because of the clear geographical association with the 1957 Windscale accident, but also for other reasons, this issue became widely debated. The Inquiry itself, although specific to this particular planning development, provided a platform for a very wide-ranging coverage of the issues involved (Parker, 1978). There were many witnesses called, and a wealth of expert opinion was expressed. An important component was the risk-assessment study that had been carried out to estimate the likelihood and consequences of failure of various items of the reprocessing plant. Once again expert technical opinion was seen to be one of the components at the centre of the issue.

The Canvey Report
In mid-1978 the Health and Safety Executive published the report of a major technical risk assessment concerned with the large chemical industry complex at Canvey Island (Health and Safety Executive, 1978). As in the case of the Windscale Inquiry, a planning application had been made to build a potentially hazardous plant on a particular site. The planning inquiry inspector had recommended that a technical risk assessment be performed and the Safety and Reliability Directorate of the UKAEA were retained as the technical consultants to carry this out. There has been widespread subsequent discussion of the report and Cremer and Warner have published a critique (Cremer and Warner, 1980) of the study. The Cremer and Warner critique is important in that it reveals once again the fact that technical risk assessment contains many topics where informed guesswork is the best that can be done in the short term to

circumvent certain areas of uncertainty. None the less, at every stage in the Canvey Island issue the role of the technical expert has been highly influential.

The above review should serve to illustrate the fact that a number of major issues in technological risk have given the appearance of being highly influenced by technical matters. It is not surprising to discover that many people interpret this to mean that technologists are offering their way of thinking as a means of settling the questions involved, offering as it were a set of mathematical formulae by means of which acceptability of risk can be rationally determined. I certainly do not believe that such a resolution is possible, however appealing it may be, but the possibility that such a scheme could be devised was perhaps what many people thought Lord Rothschild was suggesting in his widely publicised Dimbleby Lecture on risk (Rothschild, 1978). The lecture contains many references to the pitfalls in comparing different expressions of risk but in the penultimate paragraph we read: 'So why not produce an index of risks, so that you can decide above what level – road fatalities, perhaps – you should get into a panic; and below what level – death from influenza – you should relax.' This statement in itself could be taken to imply that quantitative risk assessment could provide a single index for decision. There are, however, numerous factors that militate against the production of such an index and I will now explore some of these.

Limitations on the validity of quantitative risk estimates

Several of the authors in this book deal with aspects of this problem and I will not repeat what they have written. Some important general considerations are given by Dr Cohen, whilst Sir Frederick Warner gives specific examples that demonstrate the point. In what follows I deal specifically with some major problems as I understand them.

Probability
The concept of probability enters the discussion where we are concerned with the likelihood of occurrence of damaging events. There are really two aspects: first, how often may we expect failures to occur? and secondly, what is the range of possible outcomes for a given failure? In order to demonstrate the limitations of applying probability arguments in plant failure risk assessment I wish first to refer to the concept of probability for a familiar case where the meaning is exact. If I have a pair of six-spot dice then I can easily show that there are eleven possible values (two to twelve inclusive) that may be obtained through the thirty-six possible combinations of faces. There is only one way of throwing double-six, so the probability is expressed as 1/36; there are six ways in which a seven can be displayed, so the probability is 6/36 or 1/6. These are very familiar arguments, and the probability estimate is widely understood to mean that if I throw the dice many times then on average 1/6 of the total throws will yield a seven. This brings us to the first difficulty. What validity does the probability estimate have if I throw the dice once only? In risk assessment we often have to produce estimates for events of very low frequency, and there is little reason to suppose that a person would make the same 'rational' bet if only one throw is to be made. On a large number of throws I have good confidence that one-sixth of them will be sevens and will bet accordingly. However, if the wager is to be on one throw the bargain is different. I may then decide my wager not on the one-in-six probability that a seven will be thrown but on the *possibility* that a double-six will be thrown, without regard to the quantitative probability estimate.

The application in risk assessment and acceptability is

that for low-frequency events the probability estimate is not based on a large number of trials and the public evaluation may well be more conditioned by how bad the outcome might be, with little regard for arguments as to how unlikely it is.

The dice throw analogy probably needs some modification to allow for misthrows but the features that give acceptable meaning to the probability estimates are:

(*a*) The number of possible outcomes is fixed and known; the number of faces on the dice is known.

(*b*) The consequence of each outcome is of known magnitude; the numbers on the dice faces are known.

(*c*) The likelihood of valid throws and of misthrows can be determined by experiment; the system can be tested to verify the probability model.

Given these features a credible and complete analysis can be done; knowing in addition how often the dice are to be thrown I can give a figure for the frequency of occurrence of given outcomes.

The situation is very different if I wish to estimate the consequence and probability combinations for an industrial plant system:

(*a*) Application of event and fault tree-analysis cannot ever be shown to represent every possible outcome; there can always be some failure sequence that has not been allowed for. This is like throwing a pair of dice with an unknown number of faces.

(*b*) The magnitude of the consequence cannot be exactly calculated; various sources of uncertainty enter the problem. This is like throwing dice where the numbers on the faces are not exact but are thought to be within certain ranges. The number actually on a particular face is not revealed until the throw is made.

(*c*) Only certain elements of the system can be experi-

mented with to test the model. Reliability data for
some components may be available, but other
necessary information is inaccessible. This is like
throwing dice where only some of the numbers on
the faces are known in advance.

This shows that the concept of probability cannot be
given a precise meaning in this context.

*Discrepancies between estimates of damage and the
historical accident record*
If one compares the experts' best estimates of the consequences of an accident with the historical record it is
often found that the estimates are greatly in excess of
the consequences actually manifested. This problem is a
major one and divides those responsible for safe plant-
operation from those who carry out estimates of risk.
For example, the Canvey Report contains an estimate
(p. 123) that a rapid release of about 20 tonnes of
chlorine gas could possibly kill up to 120 people in a
typical rural area in the UK and up to 6000 people in an
urban area. Expressed on a per-tonne-released basis that
is 6 deaths per tonne (rural) and 300 per tonne (urban).
If we look at the record of chlorine accidents this
century (HSC, 1979) we find 112 deaths incurred in 17
incidents releasing a total of 361 tonnes, an average of
0.3 deaths per tonne. Furthermore, none of those deaths
occurred in the UK.

There are numerous factors that may tend to reduce
the hazards from airborne toxic clouds, as detailed in
appendix 3 of the Canvey Report, but this kind of dis-
crepancy calls into question the basis of acceptability
decisions based on risk comparisons. Should we take
more note of the estimates which refer to the potential
for harm, but which may be overly pessimistic, or
should we believe that the historical record is a better
indicator, even though it may not be a good representation of what could happen? This is an issue that divides

technical experts, depending on their inclinations; perhaps a middle road is called for, but this adds to the difficulties of establishing criteria.

The validity of intercomparisons
The basis for comparison between different ways of expressing risk and between different risks expressed in the same way is a subject that I address again in this book, but there is a particular problem, concerning the unknown degree of inclusion of human factors in risk quantifiers, that needs special mention. It is well recognised that event and fault tree-analysis as practised in the context of risk assessment cannot adequately identify or include the contribution of operator error. The approach to safety in potentially hazardous systems includes three elements:

(*a*)   Reduction of the failure susceptibility of the hardware by attention to design.
(*b*)   Reduction of the susceptibility to operator error by attention to training.
(*c*)   System elements and procedures to mitigate the consequences of failure in (*a*) and (*b*).

Although the contribution of human (operator or design) error is recognised it is not readily quantifiable. Current ability to model human behaviour is not in any sense comparable to that possible for physical/mechanical/ chemical systems. Thus the *a priori* estimates of risk expressed in the Canvey Report or the Rasmussen Report cannot be thought of as comprehensive. Human error may be included in some elements, but to an unknown degree.

If, as appears now to be thought in the case of the Three Mile Island accident, operator error can be significant compared to system failure this complicates the basis of comparison between risk estimates based on a system-failure analysis and revealed risk as obtained from the historical record, since the latter form includes the human error element implicitly and completely. For example, operator error is almost certainly the main cause of road accidents and one would need in some way to disaggregate this source of risk from the others involved to obtain a fair comparison with an estimate based more squarely on the likelihood of system faults. This, too, militates against the likelihood that a single valid index of risk could be established.

Hindrances to resolution of the risk question

As one who has come into the risk business from the technical assessment sector I recognise a number of factors that seem to me to be considerable pitfalls and which do not assist in resolving the problem. For the most part these arise from misconceptions about the nature of expert technical opinion.

First, there is the unproductive competition between 'objective' and 'subjective' as labels to attach to engineering science and social science activities respectively. This is a wasteful competition that could be easily resolved by recognising that for the most part the participants intend only to distinguish between what is experimentally reproducible within certain limits of uncertainty and what is either unknown or capricious. The use of 'objective' and 'subjective' almost as perjorative terms is counter-productive. It would be very sad indeed if worthwhile discussion were to be hindered by being diverted into what appears to be an unresolvable issue in semantics.

Secondly, if those who put forward ways of expressing risk do not take good care to ensure that the limitations are made explicit then there is a powerful danger that people will interpret the willingness of the technical expert to give quantitative estimates as signifying an underlying ability to reduce risks to any desired level, given sufficient expenditure. It should be appreciated that there are uncertainties and probability distributions that are irreducible, e.g. the probability distribution of consequences arising from the effect of different weather conditions on the dispersion and uptake of toxic materials.

Thirdly, there is the danger of proposing risk criteria that do not take into consideration the importance of different perceptions of risk. Risk perception, like risk itself and like public opinion, is diverse and disaggregated. There seems little virtue in proposing controversial criteria that cause public disquiet about features that are probably not representative of the real risk involved with a particular system. Equally, if significant factors (e.g. human error) are hidden or ignored in the expression of risk then those criteria are unlikely to be found acceptable.

Fourthly, there is the danger associated with the unfathomable process whereby proposals for design guidelines become progressively and unintentionally transformed into firm criteria of acceptability of risk. This transformation can be illustrated in the case of the Farmer Curve, which is simply a release-frequency limit line expressing the idea that accidental releases of iodine—131 from thermal nuclear reactors should be constrained to occur with a decreasing frequency as the quantity released increases. In Farmer's original paper (Farmer, 1967) a graph was presented showing a line for which the product of frequency and quantity released was constant. He then went on to state 'all parallel lines of equal slope join points of equal risk in terms of curies per year. One such line might be used as a safety criterion by defining an upper boundary of permissible probability for all fault consequences.'

The Farmer line, as expressed originally, is thus a possible form of design criterion in terms of releases of radioactive material, but not in terms of casualties amongst the population. Beattie, Bell & Edwards (1969) extended Farmer's treatment by interpreting the conse-

quences of releases on the Farmer limit line in terms of
the number of casualties, which depends of course on
the population distribution around the site. It is clear,
therefore, that the original Farmer line did not carry
with it any quantitative statement about *acceptability*,
but it did have implications concerning the magnitude
of consequences. By 1976 it was possible for Farmer &
Beattie to write 'The line is regarded as separating an
upper area of unacceptably high risk from one of lower
and acceptable risk . . .'. In 1978 Clarke & MacDonald
refer to the line as defining 'a continuous boundary
between acceptable and unacceptable accidents in terms
of their estimated frequencies of occurrence . . .'.

Whatever the intentions of these authors, it is certainly
possible to understand from these words that the
Farmer line had by this time acquired the status of a
criterion of acceptability for accidents, even though the
consequences in terms of casualties are dependent upon
population distribution. One cannot find any evidence
demonstrating that this apparent transformation was
legitimised by any process whatsoever, but it is easy to
see how the form of words expressing the design criterion has acquired an air of legitimisation over the years.
This is not to say that any deliberate manipulation has
occurred, but rather to point out how subtle changes in
the forms of words used by different authors have
endowed the concept with a value judgement that goes
beyond the scope of Farmer's original statement.

Conclusion

The final concern of any process in the management,
assessment and acceptability of risk is to answer the
following question: given that technological enterprises
involve the potential for harm, which ones should we
undertake and how should we spend our safety money?
Most people involved think they know what risk is, but

their view may be coloured by the particular aspects that they choose to be involved with. The purpose of this book is to bring together many different views of the problem. By synthesising these different views of risk it may be possible to enlarge the understanding of the whole issue.

References

Factory Inspectorate (1975). *Report of the Court of Inquiry on the Flixborough disaster.* Her Majesty's Stationery Office, London.

Atomic Energy Office (1957). *Accident at Windscale No. 1 pile on 10th October 1957.* Cmnd 302. Her Majesty's Stationery Office, London.

Beattie, J. R., Bell, J. D. & Edwards, J. E. (1969). *Methods for the evaluation of risk.* United Kingdom Atomic Energy Authority AHSB(S)R 159.

Brooks, H. (1972). Science and trans-science. *Minerva,* 10, p. 484.

Clarke, R. H. & MacDonald, H. F. (1978). Radioactive releases from nuclear installations — evaluation of accidental atmospheric discharges. In *Progress in Nuclear Energy,* 2, pp. 77—152. Pergamon Press, Oxford.

Cremer and Warner Ltd (1980). *An analysis of the Canvey report.* Oyez, London.

Farmer, F. R. (1967). Siting criteria — a new approach. *Atom,* (128), pp. 152—70.

Farmer, F. R. & Beattie, J. R. (1976). Nuclear power reactors and the evaluation of population hazards. In *Advances in nuclear science and technology,* 9, ed. E. J. Henly. Academic Press, London, New York, San Francisco.

Health and Safety Commission (1976). *Advisory Committee on Major Hazards — first report.* Her Majesty's Stationery Office, London.

Health and Safety Commission (1978). *Hazardous installations (notification and survey) regulations, 1978.* Her Majesty's Stationery Office, London.

Health and Safety Commission (1979). *Advisory Committee on Major Hazards — second report.* Her Majesty's Stationery Office, London.

Health and Safety Executive (1978). *Canvey — an investigation of potential hazards from operations in the Canvey Island/Thurrock area.* Her Majesty's Stationery Office, London.

Parker (1975). *The Windscale Inquiry, report by the Hon. Mr Justice Parker.* Her Majesty's Stationery Office, London.

President's Commission (1979). *The need for change: the accident at Three Mile Island.* Washington D.C.

Rothschild, Lord (1978). Risk (The Richard Dimbleby Lecture). *The Listener,* 30 November 1978.

United States Nuclear Regulatory Commission (1975). *Reactor safety study.* USNRC, Washington D.C.

2 The nature of decisions in risk management

A. V. Cohen

Introduction

I shall try to stick close to my title. What are the risks, how are they (and how should they) be managed, who decides on the risks, and by what criteria, and how do the decision makers (whoever they may be) account to society for their decisions?

The views I express must be personal, and do not necessarily coincide with the views and policy of the Health and Safety Executive, although I am grateful to my colleagues with whom I have discussed these issues. These views will be in general terms and cover risks which are wider than the formal responsibilities of the Health and Safety Executive. So no doubt those people who have other specific responsibilities in risk management will excuse me if my general remarks have bearing on their work too.

Risk assessment is, of course, a substantial and growing aspect of engineering, and 'risk acceptability' studies are beginning to produce a literature of their own. The term 'risk acceptability' seems to be generally adopted, although what is at stake is of course the point at which a risk is perceived to be *un*acceptable. Some useful books on the subject are those by Lowrance (1976), The Council for Science and Society (1977) and Rowe (1977). These are sometimes rather theoretical, and one of them could be read to imply that a calculus could be constructed to determine risk acceptability: a point to which I shall return.

Is there Absolute Safety? The meaning of 'safe'

The term 'safe' is often used in the absolute sense, but absolute safety must be rare. Even houses are not absolutely safe! The normal use of the term to cover industrial systems and processes implies design, construction and operation to accepted standards embodying current good practice, and, in the opinion of those drafting the standard, 'safe' — a term not normally expressed quantitatively. Yet an engineering system may have about it some latent and unsuspected fault of design, and most systems are open to some lapse of construction standard or operation procedure. Or some advance in public health knowledge may suggest that some waste product, hitherto regarded as safe, needs more careful treatment. In these cases the standards are tightened (perhaps after a Court of Enquiry if some catastrophe has happened) and the system continues to be described as 'safe'.

Complex industrial installations present a further aspect of this: the potential for serious hazards means that the kind of problem that might arise needs to be thought through very systematically. Possible dangers, of the kind which would normally arise only once in a hundred, or even a thousand years, and then be the subject of a searching Enquiry, have to be anticipated and the necessary precautions taken in advance. The result of a formal fault analysis of this kind can only be a statement that (to the best of the analyst's belief) the chance of a catastrophe which might affect m people is 10^{-n}. This may seem more dangerous than the customary assurances of safety: but (provided the analyst has thought of everything!) may well actually represent a much safer system.

Thus safety is a relative term. It may in many circumstances have to be bought at a price. This has for a long time been recognised in British practice by our normal standard of safety, as expressed in the Health and Safety at Work Act and the Factories Act; 'so far as reasonably practicable'. This term has been defined in the Courts. The formal definition given by *Redgrave* is:

'Reasonably practicable' is a narrower term than 'physically possible', and implies that a computation must be made in which the quantum of risk is placed in one scale and the sacrifice involved in the measures necessary for averting the risk (whether in money, time or trouble) is placed in the other, and that, if it be shown that there is a gross disproportion between them — the risk being insignificant in relation to the sacrifice — the defendants discharge the onus upon them. Moreover, this computation falls to be made by the owner at a point of time anterior to the accident. (Fife & Machin, 1976.)

But some other aspects of safety demand the stricter standard of 'practicable' and others seem to demand an absolute duty. Thus typically, in the Factories Act, dangerous parts of machinery 'shall be securely fenced'. But even this apparently absolute duty is moderated by the Courts' criterion of 'reasonable foreseeability': although a much stricter standard than 'practicable', it implies that even here absolute safety may not be achievable.

The lawyers may wish to say something on this. But some interesting questions are relevant. If the 'reasonably practicable' is required, someone has to decide 'anterior to the accident' what is reasonably practicable, and standard setters have to have similar issues in mind. And, at the strategic level, someone has to decide whether to insist on the 'reasonably practicable' or some

stricter standard. In the example I have quoted, it is
Parliament in the Factories Act. Under the Health and
Safety at Work Act regulating power rests with the
Secretary of State acting on the proposals of the Health
and Safety Commission.

The general point embedded in this detailed discussion is that safety is not absolute, but may have to be
bought at a price, that quasi-economic criteria become
important in most circumstances, and that even where
stricter standards than those implied by 'reasonably
practicable' or something like it, are indicated, a conscious decision is needed to require those stricter
standards. And something like this must surely apply to
all risk managers, not just HSE. Some judgement of
'risk acceptability' is implicit in any legislation.

Risk decisions by the individual

Turning now to decisions by the individual as to what
he should tolerate as risk, it seems ethical that no
imposed risk can be acceptable unless:

(*a*)   Some corresponding benefit occurs to the indivi-
       dual at risk, and
(*b*)   Everything reasonable (in whatever sense is applied
       to the word) is done to reduce it, so that
(*c*)   The individual then judges that he has a good
       bargain.

This seems fine in principle, but even at this stage quasi-
legal concepts like 'reasonable' are inevitable. And the
individual may not have perfect knowledge of the
nature of the risk: indeed perfect knowledge may not
exist anywhere. He needs to know too what benefit he
enjoys, and the implication is that a higher level of risk
may be tolerable to the individual in the work-place
than where exposure is involuntary. Where both risk and
benefit are purely personal and voluntary (smoking,

dangerous sports), this may seem a matter purely for the individual. But even here, society may have a legitimate voice — the activity may injure or outrage others.

The structure of risk decisions

In most instances, however, society has a much stronger interest. Typically, an activity which benefits some may generate risks to others, and ordinary market forces may not ensure an equitable redistribution of risks and benefits. So the problem becomes social and political: one needs to ask oneself clearly at each stage who is judging acceptability and unacceptability, and by what criteria.

What emerges from all this is that there seem three stages to decisions on risk. While most of what follows will be about social or governmental decisions, it is interesting that these three stages exist even at the individual level:

(a) *Risk assessment.* How big is the risk? How well can we assess it? Can we do anything to reduce it? If so, at what cost? What benefit flows from the activity?

(b) *The actual decision.* Should we put up with the risk, insist on stricter standards or controls, or prohibit altogether? And whose voices should be heard in this decision? These are essentially judgements on acceptability or unacceptability.

(c) *'Legitimisation' of the decision.* Are decisions, either individual or corporate, acceptable to society as a whole?

This structuring is one of convenience and of function, but there are overlaps. Risk assessment is essentially technical, though it does not always lead to absolute answers. Judgements of 'acceptability' are for the 'decision maker', whoever he may be. He has to bear in mind both technical assessments and public reactions.

'Legitimisation' is a matter of democratic process, public accountability, and so on. While risk assessment should in principle be the same world-wide, judgements on acceptability, and still more the process of 'legitimisation', will vary according to national practice. There is much more room for international disagreement on 'acceptability' of risk than on risk assessment, but the differences of national attitudes to acceptability can put different lights upon what in principle is an objective, but not always completely determined, assessment of risk. We will hear more on each of these points in subsequent chapters, but it may be useful to spell out some of the problems as they seem to me.

Assessment of benefit

An individual will hardly undertake any action unless he perceives a benefit, which may not always be financial. At company level the expected benefit is likely to be financial; but there will be wider social benefits best summed up in the phrase 'wealth creation'. This will include job creation, possible lower prices of products, a widened base of taxation, and even the setting up of a capital infrastructure which future generations can enjoy.

Assessment of risk

This cannot always be done absolutely. There seem at least three different kinds of risk, and there may be subdivisions:

(*a*) Risks which clearly and identifiably lead to casualties for which reliable statistics are available (fires, factory accidents, etc.).

(*b*) Those for which an effect is believed to exist but where the causal connection to the individual cannot be certain (e.g. carcinogens or radiation).

(*c*) Experts' best estimates of probabilities of catastrophes which it is hoped will never happen.

The figure put on the first kind of risk is a reasonably

reliable expression of what actually happens. But many factors must be borne in mind even here in interpretation. The figures must be presented so as to express the hazard as experienced by the group actually at risk, and not submerged in some wider population. The casualties may be the result of a series of quite different causes, and it may be necessary, for example, to distinguish immediate cause (e.g. an unguarded machine tool) from final cause (e.g. irresponsible system of work).

The second can only represent an expert's best belief. It will often be associated with some subtle aspect (behavioural symptoms, allergies) or with a time lag between initiating cause and observed effect. It may rest on epidemiological observation. It is much more likely to be firmly based when a cause-specific illness (e.g. very rare and specific form of cancer) is identified, than when it is suspected that a hazard increases marginally some widespread common illness. There are cases of long-existing dangers being suddenly identified. Where the connection between cause and effect can be established only with difficulty, how can one then define 'safe' doses? Does a formal 'threshold' protect everyone, or all but the sensitive individual, or is there no really safe level?: the latter may be true for carcinogens and radiation. In any case, it may be prudent to insist on a 'reasonably practicable' control below a formal threshold, even in other situations. There may be no way of being absolutely, or even reasonably, certain (the transscientific situation). And what does one do when a widespread practice, hitherto regarded as safe, *may* be causing a widespread effect, but this cannot be established by any normal scientific standards? There are examples of this, particularly in public health matters. National attitudes to this kind of situation may lead to wide variations in attitude to controls, and costly controls *may* be imposed whose benefit, if it exists, cannot be established.

The third problem is the classic one of a large installation with potential hazards. We are very familiar with the disagreements as to whether a nuclear power station is 'safe': similar questions arise with many industrial installations. Quite apart from the question of whether a predicted level of probability is 'acceptable', there are further questions. Have the expert assessors thought of everything that can go wrong? With what confidence can a probability be stated? These questions are fair ones, and cause those to whom the risk is, in any case, unacceptable to call in question (often very unfairly) whether the expert assessment is fair and unbiased.

The second and third categories imply uncertainty, and often deal with 'dread' hazards such as cancers, catastrophes, etc. It therefore seems unwise to boil together all predictions of consequences into some unified index of deaths or diseases — a 'unified index of woe'. A death by cancer may reasonably be regarded as different from an accident, and a 10^{-4} chance of ten thousand deaths from one death a year. Nor can we equate a death by accident as equal to six thousand people off work for a day with a cold. All these have been done by one or other author but, in my view, the hazards need to be disaggregated and displayed separately. Otherwise the decision maker will be presented with what appears to him to be a hard figure, measuring the consequences of risk; but which has concealed within it some analyst's private scale of personal values, and which makes difficult subsequent allowance for differing public levels of tolerance. Thus the uncertainties, and the 'dread' nature of the hazards, must be kept in mind by the decision maker, who will also be told by members of the public, often quite unfairly, that the expert's assessment is biased or blind to obvious risks. Real difficulties (e.g. how to cope with a suspected but unproven hazard, or a low-dose no-threshold effect) can be obscured by playing with words, or submerged within some formula which has worked well up to now. These issues complicate the process of decision and explain the reference above to overlap between the stages of risk assessment, and judgement of risk acceptability.

The actual decision
The decision maker, whoever he may be, has to decide: Should I or we put up with this hazard, assuming everything reasonable has been done to reduce it? The decision maker might be an individual, a company, a government, or a judge deciding on a civil action.

The nature of the information available differs in each case. The individual is likely to have imperfect and often self-distorted information (e.g. cigarette smoking). A large company, and still more a government, is likely to have as good information as is available, though this may be imperfect. Even at this level, judgement may be distorted, or at least seem to be so to others. And in a trial (e.g. a civil action for workman's compensation) it is possible for both sides to present expert opinions, each of which is true and consistent with scientific knowledge but which take opposite views of an uncertain situation.

The individual in judging his personal reaction to a risk is likely to have a threshold of *de minimis* perception below which he does not think about the risk, still less care: snake bites, lightning, etc. He will certainly have an upper level 'above which he will not put'. These two levels may reflect frequencies of occurrence, but will be altered by individual perception. They may vary from one to another hazard or individual. The difference between upper and lower levels, and the difference between objective risk and subjective perception, is sometimes obscured in the literature.

At company level, the decisions on safety will, or should be, determined by something much stronger than self-interest. For one thing, the employer has various duties under the Health and Safety at Work Act, and is subject to various penalties if his works do not meet required standards. Moreover, he will be conscious of the wishes of his own workforce (who also have duties under the Act, as well as obvious self-interest in pressing their safety). This meets a great deal of the criticism put by the Council for Science and Society report:

The reduction of risks requires a management prepared to invest financial and organizational resources, a workforce willing and capable of taking a positive attitude to their own health and safety, and a determined and effective inspectorate. Where such attitudes prevail, and capital is available, it can pay for itself in every way. But only too often it becomes company policy to reduce risks to workers only *after* a major public scandal, or where their hand has been forced by a determined and articulate workforce.

But of course the capital has to be available, and even if it is, any industrial organisation will wish to set priorities. But in doing so, and in attacking the more dangerous situations first, the definition of the required standard of safety must be borne in mind. It can be no defence for not controlling a hazard to the 'reasonably practicable' that priority is being given to other, and more dangerous parts of the factory! In such cases, more funds will have to be spent on safety.

This is easy to say. For gross hazards, necessary pressures can be put on the occasional recalcitrant firm. But at national level, decisions have to be made on how far to insist on controls of grey areas, and how rapidly to insist on controls which advancing knowledge show necessary, without bankrupting the country. Something analagous at national level to the individual balancing of risks and benefits must be carried out. But how is this to be done? Can it be done by some integration of individual perceptions? And, if so, how does one discount the individual extreme view, or the wide variations in level of perception? And how does one decide whether to control to 'reasonably practicable' or to a stricter standard, or to prohibit altogether?

One hypothesis, which is purely personal, is that a judgement, not always explicit, is formed by the decision maker whether the overall or average level of un-acceptability lies above or below the threshold of perception. If it lies above, then control to 'reasonably practicable' in between the two levels is the right procedure. If it does not, then one either has to control, probably more strictly than to 'reasonably practicable', to the level below which no risk is perceived (the *de minimis* level) or prohibit altogether if this seems unacceptable, and if substitutes are available (white phosphorus, certain very potent carcinogens). There may genuinely be no substitute for a hazardous substance or process which causes real concern. Such judgements must be essentially political in nature.

Attempts have been made by some authors to formalise the ways in which the public differentially perceives risk. A well known and perhaps the most realistic, attempt has been made by Lowrance, in qualitative terms, in a table which speaks for itself (Table 1).

What is more questionable is any attempt by others to assign numerical weights to such differential aversions. If this could be done, then in principle one could, so to speak, arrange risk control resources to minimise on public anxiety rather than on objective risk (and incidentally sub-optimise on the latter). Put this way, this would seem an ill-advised attempt to by-pass the political process. But if there is any reasonable uniformity or

Table 1.

Risk assumed voluntarily	Risk borne involuntarily
Effect immediate	Effect delayed
No alternatives available	Many alternatives available
Risk known with certainty	Risk not known
Exposure is an essential	Exposure is a luxury
Encountered occupationally	Encountered non-occupationally
Common hazard	'Dread' hazard
Affects average people	Affects especially sensitive people
Will be used as intended	Likely to be misused
Consequences reversible	Consequences irreversible

consensus in differential public aversions, and if these
do not change in time — and these are two big ifs — then
some calculus along these lines might be useful in
ensuring that reasonably consistent decisions are made
by the decison makers uninfluenced by temporary particular pressures.

Such a calculus would need to be used with caution.
There is a danger otherwise that one man's set of utilities
will be buried in submissions to decision makers and lost
sight of. Typical of such dangers are attempts to put a
value on life (the decision maker is not always reminded
whether this includes or excludes subjective judgements
of intrinsic values) or 'objective' assessments of risk
which include 'unified indices of woe' of the kind I
referred to earlier. Hidden values can present real
problems in other areas too. What weight should be
given at the strategic level to economic criteria (do we
shut a firm down or tolerate a marginally greater risk for
a bit longer?). To what extent do differential perceptions
of risk act as an effective tax on technological innovation? These are essentially political judgements, which
must not and cannot be replaced by a calculus.

The process of decision making at this level is thus
essentially political. The expert has a role in sitting in to
ensure that his expert views are taken account of and
not misunderstood on the way up, and to contribute his
own common sense as an official — but his scientific
background here is to be regarded as background to his
official task. In thus acting, the scientist has a duty of
conscience too, as summarised by Lowrance, in such
matters as advising where 'technical components need
to be distinguished explicitly' or where risks 'appear so
grave or irreversible that prudence dictates the urging
of extreme caution even before the risks are known
precisely'. But in all his contributions to this essentially
political process the scientist's views cannot over-ride
others' — wider issues may be the determining ones.

Legitimisation of the decision

However the decision is made, it must be conducted with absolute integrity. Society must remain confident of the way government is acting for them, otherwise accusations of interested bias will be made. This is, of course, the reason for the 'legitimisation' procedure and for increasing transparency of government. This is a public issue which runs much wider than safety matters, and each nation will have its own way of dealing with 'legitimisation'. We, of course, have Ministers in Parliament, the representative nature of the Health and Safety Commission, and the consultative process: Green Papers, consultative draft regulations, public hearings, appeal procedures, and so on. The extent of this kind of procedure has increased in recent years, presumably in response to changing public need. There are some who maintain that greater transparency of the decision process is needed. This, too, is a matter of political judgement.

Other nations proceed in different ways: many of our continental neighbours find the consultative process even newer than we do; and the Americans are used to decisions themselves being made in a public adversarial situation. The value of 'legitimisation' is something more than instilling a sense of 'participation'. Harvey Brooks (1972) has remarked 'Adversary procedures may be especially valuable in bringing out unanalyzed evaluative assumptions or premises which underlie the testimony of experts when they deal with trans-scientific issues.' While this refers explicitly to the American adversarial procedures, it is of wider application, and I hope that the sociologists will be able to advise us what, in any one country, constitutes a desirable procedure. I am sure that there is no absolute formula and much depends on national traditions.

Conclusions

Some of the problems may seem very familiar to lawyers and sociologists and, if so, I apologise for appearing to rediscover the wheel! I suggest, though, that the uncertain nature of risk management raises questions which normally do not arise elsewhere, and which put a new aspect on issues which seem familiar to them, just as their outlook enables us to realise latent problems in what hitherto has seemed easy. To illustrate this, I will end with some questions:

(*a*) How does one disaggregate statistics and estimates of risk so as best to help the subsequent decision process?

(*b*) How does one ensure that the public are aware of social benefits from a project, as well as the risks?

(*c*) Where should the cost of control lie? Does the 'Polluter Pays' Principle lead to transfer of cost to the consumer? If so, does it matter?

(*d*) How does one deal with real risks whose importance is not uppermost in the public mind? How does one balance these against the specific cases put by pressure groups?

(*e*) How can the legal process cope with situations of uncertainty in cause and effect (e.g. in claims for workmen's compensation)?

(*f*) To what extent do apparently unassailable forms of words contain hidden values?

(*g*) How can we avoid safety being a hidden tax on new technology?

References

Brooks, H. (1972). Science and trans-science. *Minerva,* 10, p. 484.
Council for Science and Society (1977). *The acceptability of risks.* Barry Rose, London.

Lowrance, W. W. (1976). *Of acceptable risk*. William Kaufman, Los Altos, California.

Fife, I. & E. A. Machin (1976). *Redgrave's health and safety in factories*. Butterworth, London.

Rowe, W. D. (1977). *An anatomy of risk*. Wiley, New York.

3 Benefits and risks, their assessment in relation to human needs

T. A. Kletz

'I was wondering what the mouse-trap was for', said Alice, 'It isn't very likely there would be any mice on the horse's back.'

'Not very likely, perhaps', said the Knight; 'but, if they do come, I don't choose to have them running all about.' 'You see', he went on after a pause, 'it's as well to be provided for everything. That's the reason the horse has all these anklets round his feet.'

'But what are they for?' Alice asked in a tone of great curiosity.

'To guard against the bites of sharks', the Knight replied.

Lewis Carroll, *Through the looking glass*

Introduction

Safety officers, and managers as well, often feel like the White Knight. Some risks are rather unlikely to result in anyone being killed or injured, but they might, and so they feel they must do something about them or they may be partly responsible for someone's death or injury. They feel they must be 'provided for everything'. But to remove every possible risk, however slight, is almost impossible and even if we did decide to try to do so, we cannot do everything at once. How do we decide which risks should be dealt with first, which can be left, at least for the time being? In short, how do we allocate our resources?

In the past this dilemma was often resolved by spending lavishly to remove all possible risk from those hazards which had been brought to our attention by an accident and ignoring the rest. It was sometimes necessary to avoid looking too hard for hazards in case we found more than we could deal with. Such a method is, of course, wrong. Whether resources are large or small we should spend them in a way which maximises the benefit to our fellow men and which can be logically defended. The latter is important. If I say that the risk from lightning, the transport of chemicals, or rock-climbing is small and should be ignored, while you say it is high and demands immediate attention, discussion between us is difficult. If, however, there is an agreed scale for measuring risk, a dialogue becomes possible.

First, we should bear in mind that we are all at risk all the time, whatever we do, even if we stay at home doing nothing. We accept risks when we consider that by doing so something worthwhile is achieved. We go rock-climbing or sailing or we smoke because we consider the pleasure worth the risk. We take jobs as airline pilots or soldiers or we become missionaries among cannibals because we consider that the pay, or the interest of the job, or the benefit it brings others, makes the risk worthwhile.

At work, whatever we do to remove known risks, there is likely to be some risk, however slight, to employees and to members of the public who live nearby. By accepting this risk we earn our living and we make goods that enable ourselves and others to lead a fuller life.

Different numerical approaches

During the last ten or fifteen years attempts have therefore been made to apply numerical methods to safety problems. These attempts have used one of two distinct methods.

(a) Weighing in the balance
In this method, sometimes known as the trade-off, the benefits and disadvantages of various courses of action are expressed in common units, usually money, so that they can be offset against each other and the course of action giving the biggest net benefit identified. We weigh in the balance the pros and cons of each proposal.

For example, we can compare the cost of preventing an accident with the costs of the damage and injury it will produce, multiplied by the probability that it will occur, or we can compare the cost of preventing pollution with the damage caused by pollution.

In the UK this approach is sanctified by the law. The words 'reasonably practicable' which occur so often in our safety legislation have been defined as implying

that a computation must be made in which the quantum of risk is placed in one scale and the sacrifice involved in the measures necessary for averting the risk (whether in money, time or trouble) is placed in the other, and that, if it be shown that there is a gross disproportion between them — the risk being insignificant in relation to the sacrifice — the defendants discharge the onus on them. (Fife & Machin, 1976.)

Weighing in the balance is satisfactory if we are considering accidents that could cause damage to plant and loss of production, but are unlikely to injure anyone. However, when we consider accidents that may kill or injure people, despite what the law says, it becomes more difficult, because we have to put a value on human life. I am aware of the many courageous attempts (Jones-Lee, 1976; Mooney, 1977) that have been made to do so, but they have not been altogether accepted, and it is desirable to avoid using these figures if we can.

Similarly, attempts to put money values on the intangible effects of pollution have proved difficult. Overall balances have been achieved in only a few cases

(Frost, 1971; Programmes Analysis Unit, 1972) and the results have not been widely accepted.

For these reasons most of the attempts to apply numerical methods to safety problems have used the second method, standard or target setting. This is the method used by the law in the UK in practice if not in principle.

(b) Standard or target setting

A company, industry or government sets a standard which must not be exceeded or a target which should be aimed for. Such standards or targets cover widely diverse hazards, for example, they specify the height of handrails, the concentrations of toxic chemicals in the atmosphere, the level of noise or the amount of pollutants that can be discharged to the atmosphere or a river. Ideally, such targets should be set so that the levels of risk are comparable; we should not spend money on raising the height of handrails if the risk of falling over them is smaller than the risk of being harmed by toxic chemicals in the atmosphere. In practice, of course, the targets used are often out of line, even within the same industry.

I would now like to describe some methods that have been developed, using this second approach, to help us allocate our resources rationally in safety matters. They have been developed mainly to deal with problems arising from acute hazards such as fires, explosions and large releases of toxic gas, but attempts are now being made to apply them to chronic hazards. The justification of these methods is not, of course, that they are philosophically sound (though one would like to think that they are) but that they help people in industry to use their resources more effectively to the greater benefit of their fellow men. They are essentially designed for use by practical people. Further details are given in

Kletz (1971, 1972, 1976*a*, 1977, 1978), Bulloch (1975)
and Gibson (1976).

Risks to employees

Before we can set a target for safety we need a scale for
measuring it. One such scale is the fatal accident frequency rate (FAFR), the number of fatal accidents in a
group of 1000 men in a working lifetime (100 million
man-hours). The British chemical industry's FAFR is
about 4 excluding Flixborough, or about 5 if Flixborough
is averaged over a ten-year period.

Within the chemical industry, if we can identify an
activity which contributes more than 0.4 to the FAFR
we try to remove it as a matter of priority; lower risks
can be left for the time being. Experience has shown
that the costs of such a strategy, though often substantial, are not unbearable, despite the fact that some
competitors do not incur this expenditure. Some of the
extra costs can be recouped by the greater plant reliability which safety measures often bring; the rest is a self-
imposed 'tax' which has to be balanced by greater
efficiency.

Risks to the public

When we consider risks to the public at large from
industry, the level of risk which can be considered tolerable, even in the short term, should be much lower. A
man chooses to work for a particular employer or in a
particular industry, and, unless he chooses a particularly
hazardous occupation, the risks he runs are not much
greater than if he stayed at home. On the other hand,
the public may have risks imposed on them without
their permission, and though society as a whole may
gain, they may not. Not all the people who live near
airports wish to travel by air.

Chauncy Starr has pointed out that we accept voluntarily risks such as driving, flying and smoking, which expose us to a risk of death of one in 100 000 or more (sometimes a lot more) per person per year (a FAFR of 0.1) (Starr, 1969, 1972). We also accept, with little or no complaint, a number of involuntary risks which expose us to a risk of death of about one in 10 million or less per person per year (a FAFR of 0.001). Table 1 lists a number of these voluntary and involuntary risks. (The figures are, of course, only approximate and may have been calculated using different assumptions. Furthermore, some of the comparatively small number of workers in this field copy from each other, so any error, once introduced, is repeated in various papers and acquires an aura of authenticity. Any individual figure should therefore be checked in the original sources if it is used in a calculation.)

We accept very high risks voluntarily; we accept other risks, imposed on us without our leave, if they are sufficiently small. It would be possible to reduce the involuntary risks listed in Table 1 if there were sufficient pressure from the press and public, but on the whole there is no such pressure. The risk of being struck by lightning or falling aircraft is so small that we accept the occasional death without complaint. To quote the *Daily Telegraph* commenting on calls for a crash programme of snowploughs, 'This is rather like insuring oneself against snake-bite or being struck by lightning. It is impossible to take precautions against everything'. We accept very high risks travelling in road vehicles, presumably because their advantages are clear and obvious. From natural disasters we accept risks of about one in a million per person per year; from man-made events, except road transport, we seem to accept about one in 10 million per person per year.

Leukemia and influenza have been included in the list as examples of risks we do not readily accept; there is pressure for something to be done. Most people would

Table 1. Voluntary and involuntary risks compared

Voluntary		Involuntary	
Activity	Risk of death per person per year ($\times 10^{-5}$)	Activity	Risk of death per person per year ($\times 10^{-7}$)
Smoking (20 cigarettes per day)	500	Run over by road vehicle (USA)	500
Drinking (1 bottle wine per day)	75	Run over by road vehicle (UK)	450
Football	4	Floods (USA)	22
Car racing	120	Earthquake (California)	17
Rock-climbing	4	Tornadoes (Mid-West USA)	22
Car driving	17	Storms (USA)	8
Motorcycling	200	Lightning (UK)	1
Taking contraceptive pills	2	Falling aircraft (USA)	1
Taking saccharin (average US consumption)	0.2	Falling aircraft (UK)	0.2
Eating peanut butter (4 tablespoonfuls per day)	4	Explosion of pressure vessel (USA)	0.5
Diagnostic X-rays (average US exposure)	1	Release from atomic power-station (at site boundary) (USA)	1
Being in same room as smoker (average US exposure)	1	(at 1 km) (UK)	1
		Flooding of dikes (Holland)	1
		Bites & stings of venomous creatures (UK)	1
		Transport of petrol & chemicals (USA)	0.5
		(UK)	0.2
		Leukemia	800
		Influenza	2000
		Meteorite	6×10^{-11}
		Cosmic rays from explosion of supernovae	$10^{-8} - 10^{-11}$

Data mainly from Starr (1969, 1972), Pochin (1973, 1975), Gibson (1976) and Hutt (1978).

The last four figures on the left-hand side are estimated by Hutt assuming a linear response and could be wrong if there is a threshold exposure below which no harmful effects occur.

support action to reduce the incidence of these diseases, but would regard the others as hardly worth bothering about.

We thus have a basis for assessing risks to the public at large from an industrial activity. If the average risk to those exposed is less than one in 10 million per person per year, it should be accepted, at least in the short term, and resources should not be allocated to its reduction. A risk of one in 10 million per person per year is, of course, extremely low. It may be easier to grasp if it is expressed as follows: Suppose all sources of death were removed excepting that resulting from a particular industrial activity, then all the people living near the factories concerned would live, on average, for 10 million years!

This risk is a good deal lower than that proposed in the Canvey Island report (Health and Safety Commission, 1978a). However, the report, in many people's view, has exaggerated the size of the risks, and so the difference is not as great as it seems at first sight. The report admitted on the last page that it may have exaggerated the risks, as it stated, 'Practical people dealing with industrial hazards tend to "feel in their bones" that something is wrong with risk estimates as developed in the body of the report'. Nevertheless, the report is a landmark as it shows official acceptance of the view that we cannot do everything possible to avoid every conceivable accident, and that numerical methods should be used to decide what level of risk to accept.

This brings me to the use of the term 'acceptable risk' which many people find repugnant, and I can sympathise with their view. 'We should never', they say, 'deliberately accept risks to other people'. Of course, we should not, but we cannot do everything at once; some things have to be done first, others left until later. Hazard analysis, to repeat what has been said earlier, is concerned with priorities rather than principles.

Examples

I would like to give some examples of problems to which numerical methods and the criteria described in this paper or similar criteria have been applied in order to decide whether or not we act to reduce the level of risk, or protect people from the consequences is justifiable at the present time. Most of my examples concern risks to the public but there are many examples of risks to employees in the literature (Kletz, 1971, 1972, 1976a, 1977, 1978; Bulloch, 1975; Gibson, 1976).

(a) Escapes of toxic gases
Dicken (1974) and Sellers (1976) have described a method for identifying all the circumstances which could lead to an emission of chlorine from a plant and the size and probability of a release. The concentration of chlorine at the plant boundary is then estimated and compared with target figures. A 'nuisance' is considered acceptable once a year, a release causing 'some distress' is considered acceptable once in ten years, and a release which could lead to 'personal injury or risk to life' is considered acceptable once in a hundred years. All these expressions are quantified in terms of concentration and duration. It can be shown that the last category is roughly equivalent to a risk of 10^{-7} per person per year for the population living near the plant.

Siccama, of the Directoraat-Generaal van der Arbeid, the Dutch Factory Inspectorate, has discussed the risk to the public from the storage of acrylonitrile (Siccama, 1973). He estimates that, if a tank is situated 2500 m from a residential area, the general public will suffer 'irreversible negative effects' once in 60 000 years. If there are six tanks in a group, the effects will be suffered once in 10 000 years, which Siccama considers acceptable.

The figure of once in 10 000 years is, in fact, the frequency with which dikes in Holland are liable to be flooded and results, as shown in Table 1, in a risk of death for the people living behind the dikes of 10^{-7} per person per year. The argument, presumably, is that if resources are not being spent to reduce the risk of flooding and drowning below a certain level, why should the risk of acrylonitrile escaping and gassing people be made smaller? Siccama's paper is interesting as another example of the use of a quantitative approach by an official body.

Toxic gas releases are also discussed in the Canvey Island report where large numbers of casualties are considered possible, though unlikely. The experience of the industry, however, is that when toxic gases have been released, the number of casualties have been relatively small, about 1 per 3 tonnes for chlorine, 1 per 50 tonnes for ammonia (Advisory Committee on Major Hazards, 1979).

(b) Unconfined vapour cloud explosions

I have shown (Kletz, 1978) that in many plants handling large quantities of hazardous materials the chance of a leak followed by an unconfined vapour cloud explosion is not so low that it can be ignored, and that some strengthening of control rooms is desirable. I have also shown (Kletz, 1979) that if plants are located so that the overpressure developed at the nearest houses is less than 0.7 psi, then the chance that members of the public will be killed is less than 10^{-7} per person per year.

(c) Pipelines

An estimate of the probability that a cross-country pipeline will be damaged and that the public will be 'affected', is given in the Health and Safety Commission (1978b) report. If 'affect' is assumed to mean a 1 in 430 chance of death, then the level of risk to the population is similar to that suggested in this paper.

(d) Industrial hygiene

Pochin (1973, 1975) has compared deaths from in-

Table 2. Estimated rates of fatality (or incidence) of disease attributed to types of chemical or physical exposure

Occupation	Cause of fatality	FAFR
Shoe industry (press & finishing rooms)	Nasal cancer	6.5
Printing trade workers	Cancer of the lung and bronchus	10
Workers with cutting oils:		
Birmingham	Cancer of the scrotum	3
Arve District (France)	Cancer of the scrotum	20
Wood machinists	Nasal cancer	35
Uranium mining	Cancer of the lung	70
Coal carbonisers	Bronchitis & cancer of the bronchus	140
Viscose spinners (ages 45–64)	Coronary heart disease (excess)	150
Asbestos workers:		
Males, smokers	Cancer of the lung	115
Females, smokers	Cancer of the lung	205
Rubber mill workers	Cancer of the bladder	325
Mustard gas manufacturing (Japan 1929–45)	Cancer of the bronchus	520
Cadmium workers	Cancer of the prostate (incidence values)	700
	Asbestosis	205
Amosite asbestos factory	Cancer of the lung & pleura	460
Nickel workers (employed before 1925)	Cancer of the nasal sinus	330
	Cancer of the lung	775
β-Naphthylamine manufacturing	Cancer of the bladder	1200

dustrial disease with those due to industrial accident. Table 2 is taken from Pochin (1973) (the units have been converted from deaths per million men per year to FAFR).

Many of these figures are high compared with accident rates, and would justify high priority and urgent action. However, because industrial disease takes many years to develop, they may refer to past working conditions rather than present ones. The figures do suggest, however, that until we are confident that the figures have been reduced, improvement to working conditions in these industries justifies high priorities.

The legally permitted levels of asbestos in the working atmosphere in the UK have not been set so that there will be zero evidence of asbestosis in those exposed, but so that less than 1% of those exposed for a working lifetime will show clinical signs of a disease (Roach, 1976). Here we have an official admission that a compromise is necessary between cost and safety, but there has been no attempt to relate the level of safety to that achieved elsewhere.

Similarly, exposure to the threshold limit value (TLV) concentration of a chemical in the working environment does not mean that the chance of harmful effects is zero. According to Roach (1977), for those chemicals — a minority — for which the threshold limit value is based on a risk of fatal consequences, exposure to the TLV for a working lifetime produces a 1 in 10 000 chance of death, equivalent to an FAFR of 0.1. This suggests that TLVs are based on a level of safety comparable with that which is aimed for generally in the process industries. Expenditure to achieve a TLV is reasonable; expenditure to go below is not reasonable at the present time.

However, Roach qualifies his estimate by pointing out that TLVs 'have often been generated from little or, sometimes, no factual evidence', and this could result in too much priority being given to achievement of the TLVs in these cases.

(e) Human reliability

Numerical methods can be applied to problems of human reliability, indeed must be, for no system is fully automatic; there is always some human involvement, if not in operation then in maintenance and testing. If we want to know how often the system will fail, we must estimate how often the men involved will make errors or omissions, for even well-trained, well-motivated men, physically and mentally capable of doing what we ask, will make occasional mistakes. We either accept the occasional mistake − and its consequences − or we change the work situation.

It is possible to make rough estimates of likely error rates and take these into account in our calculations (Rigby, 1971; Swain, 1974; Kletz, 1976*b*).

The cost per life saved

Earlier in this paper I said that I preferred 'target setting' to 'weighing in the balance', but perhaps we should look more closely at the latter, as it may be useful as a secondary criterion. If we adopt the targets I have described, how much will we have to spend to save a life and how does this compare with the money spent elsewhere?

There are many fields in which people have to decide how much money to spend to save lives. The judgements are usually implicit. People do not consciously say, 'We will spend up to £100 000 to save a life', but nevertheless, the money allocated and the number of lives saved enable us to calculate what is actually spent. By collecting a number of such figures we should be able to find out the value society actually places on human life.

As Craig Sinclair (1972) has shown, the value placed on a life varies over a large range. In agriculture, £2000 is spent to save an employee's life, in steel handling £200 000, and in the pharmaceuticals industry £5 million.

On the whole, the newer industries spend more than the older ones. On the other hand, the pharmaceuticals industry values the lives of third parties at only £10 000.

In marked contrast, doctors can save life for, comparatively, very small sums. Gerald Leach, for example, quotes the following figures (1972 prices; Leach, 1972):

Lung X-rays for old smokers	£400
Cervical cancer screening	£1 400
Breast cancer screening	£3 000
Artificial kidney	£9 500
Isotope-power heart	£26 000

One medical writer, unaware of the money spent to save life in other fields, has written:

The total cost of a bed in an intensive care unit can be as much as £450 per week. In other words a ten-bedded unit costs nearly £250 000 a year to run, and it has been estimated that of about 500 admissions 50 lives are saved annually at a cost approaching £5000 per case. Such astronomical costs naturally raise the question as to whether it is ethical to concentrate so much resource on so small a number of patients when there are many neglected areas of medical care. (Miller, 1973.)

Although the Road Research Laboratory has estimated the value of a life, in terms of a person's future contribution to production, at no more than £15 000, in fact, much greater sums are spent on the removal of road hazards. The sums are far higher than can be justified by the direct costs of injuries and damage and imply a life valuation of up to £200 000. Jones-Lee (1976) and Mooney (1977) review the various methods suggested for valuing life. In the chemical industry, experience has shown that to achieve the levels of safety proposed in earlier sections we have to be prepared to spend up to £1 million or so per life saved.

If a particular proposal calls for the expenditure of more money than this, then it seems that we are not spending our own or the nation's resources wisely. *We should not accept the risks*; instead, we should look for a cheaper solution. Experience shows that, in practice, it can usually be found.

Once we say that risks must be removed if they are cheap to remove, but can be accepted if they are expensive to remove, then every risk may become expensive to remove. If we can say that all risks above a certain level (which may vary from industry to industry and from time to time) must be reduced, then, in practice, technologists will find a 'reasonably practicable' way of doing so.

Despite the publicity given to the single incident at Flixborough, the figures quoted earlier prompt me to ask if the chemical and process industries are too safe — do we spend too much of the nation's resources on removing the risks to employees and the public created by this sector? Would the money and time spent save more lives if some of it was used instead to take some of the risk out of coal mining or the construction industry or out of road transport? Would I be better employed in persuading people not to smoke?

Unfortunately, if the process industries spent less on safety there is no social mechanism by which the money saved could be allocated to the mines or roads. Also, society does not advance, in any field, by marching uniformly over a broad front. It improves by spearheading; one firm or industry goes out in front and shows what can be done, and the rest follow. The process industries, and many of the other newer industries, despite the intrinsic hazards of the materials they use, and despite a few incidents which have hit the headlines, have demonstrated that a high standard of industrial safety can be achieved. Perhaps the construction industry and other high-accident industries will now follow this lead. Nevertheless, some levelling out is perhaps desirable. In Britain the government controls directly both the nuclear industry and the health service, yet the implicit life valuations in these two industries are vastly different. Perhaps the nuclear industry can be restrained from spending even more money on safety and any spare funds given to the health service.

A final note

To many people, the approach of this article may seem cold-blooded and callous. I do not think it is. Safety, like everything else, can be bought — at a price. The more we spend on safety, the less we have with which to fight poverty and disease or to spend on those goods and services which make life worth living, for ourselves and others. Whatever money we make available for safety we should spend in such a way that it produces the maximum benefit to our fellow men. There is nothing humanitarian in spending lavishly to reduce one hazard because it hit the headlines last week and ignoring the rest.

Note

A large part of this paper was first published in *The Royal Society of Medicine International Congress and Symposium Series* 1980, No. 17: Human Health and Environmental Toxicants, pp. 167—76.

References

Advisory Committee on Major Hazards (1979). *Second Report*, pp. 11 and 12.

Bulloch, B. C. (1975). *Institution of Chemical Engineers symposium series no. 39a*, p. 289.

Daily Telegraph (3 January 1979). Editorial comment.

Dicken, A. N. A. (1974). *Proceedings of the chlorine bicentennial symposium*, p. 244.

Fife, I. & Machin, E. A. (1976). *Redgrave's health and safety in factories*. Butterworth, London.

Frost, M. J. (1971). *Values for money — the techniques of cost benefit analysis*. Gower Press, London.

Gibson, S. B. (1976). *Chemical engineering progress,* 72, no. 2, p. 59.

Health and Safety Executive (1978a). *Canvey — an investigation of potential hazards from operations in the Canvey Island/ Thurrock area.* Her Majesty's Stationery Office, London.

Health and Safety Executive (1978b). *A safety evaluation of the proposed St Fergus to Moss Morran natural gas liquids and St Fergus to Boddam gas pipelines.* Her Majesty's Stationery Office, London.

Hutt, P. B. (1978). Unresolved issues in the conflict between individual freedom and government control of food safety. In Conference on Public Control of Environmental Health Hazards, New York Academy of Sciences, 29 June, 1978.

Jones-Lee, M. W. (1976). *The value of life.* Martin Robertson, London.

Kletz, T. A. (1971). *Institution of Chemical Engineers symposium series no. 34,* p. 75.

Kletz, T. A. (1972). *Loss prevention,* 6, p. 15.

Kletz, T. A. (1976a). In *Chemical engineering in a changing world,* p. 397, ed. W. T. Koetsier, Elsevier, Amsterdam.

Kletz, T. A. (1976b). *Journal of Occupational Accidents,* 1, no. 1, p. 95.

Kletz, T. A. (1977). *Hydrocarbon Processing,* 56, no. 5, p. 297.

Kletz, T. A. (1978). *Chemical Engineering Progress,* 74, no. 10, p. 47.

Kletz, T. A. (1979). Plant layout and location: some methods for taking hazardous occurrences into account. American Institution of Chemical Engineers, Houston, Texas, April 1979.

Leach, G. (1972). *The biocrats,* chapter 11. Penguin Books, Rickmansworth.

Miller, H. (1973). *Medicine and society,* p. 63. Oxford University Press.

Mooney, G. H. (1977). *The valuation of human life.* Macmillan, London.

Pochin, E. E. (1973). *Proceedings of the symposium on the assessment of exposure and risk,* p. 35. Society of Occupational Medicine.

Pochin, E. E. (1975). *British Medical Bulletin,* p. 184.

Programmes Analysis Unit (1972). *An economic and technical appraisal of air pollution in the United Kingdom.* Her Majesty's Stationery Office, London.

Rigby, L. V. (1971). *Chem Tech,* December 1971, p. 712.

Roach, S. A. (1976). *Annals of Occupational Hygiene,* 13, p. 7.

Roach, S. A. (1977). *Proceedings of the Chartered Institute of Building Services summer conference, Florence, 1977,* p. B1.

Sellers, J. G. (1976). *Institution of Chemical Engineers symposium series no. 47*, p. 127.

Siccama, E. H. (1973). *De Ingenieur*, 85, no. 24, p. 502.

Sinclair, C. (1972). *Innovation and human risk.* Centre for the Study of Industrial Innovation, London.

Starr, C. (1969). *Science*, 165, p. 1232.

Starr, C. (1972). In *Perspectives on benefit—risk decision making*, p. 17. National Academy of Engineering, Washington.

Swain, A. D. (1974). *Human factors associated with prescribed action links.* Sandia Laboratories Report no. SAND 74-0051, July 1974.

4   *Problems in the use of risk criteria*

Richard F. Griffiths

Introduction

A major objective in the field of risk assessment is the development of a uniform definition of risk. Ambiguities in the expression of risk and inappropriate definitions impede the consideration of acceptability. In the next sections I will discuss a number of ways in which risk has been expressed, making some comments and proposals that I hope will contribute positively towards the development of expressions of risk that are both lucid and appropriate. The distinction between variables that are truly risk-significant and those that are merely associated but inappropriate measures of activity will, I hope, be made clearer.

Individual risk

This is one of the most widely used measures of risk and is simply defined as the fraction of the exposed population suffering a specific effect per unit time. Thus, from data on causes of death in the UK population, Grist

(1978) has compiled a comprehensive inventory of individual risk for various natural and accidental causes. Approximately 640 000 deaths occur each year in the current UK population of 54.4 million, from all causes, and this yields a figure of 1.2×10^{-2} per year as the total individual risk of death averaged over the whole UK population. It is instructive to note that accidental death constitutes only 2.8% of the risk from all causes, and that about two-thirds of the overall accidental death risk arises from road accidents and falls. The classification can be broken down in various ways, e.g. according to age (Table 1) revealing that the individual risk from all causes is high in the first four years of life (3.3×10^{-3} per year) but drops to one-tenth of this value for the 5 to 9 age group. Thereon it increases steadily such that individuals in their mid-forties are again subject to a risk of death of about 3×10^{-3} per year. At age 70 the risk is 10 times this level (3×10^{-2} per year), whilst for the 85+ age group the level is about 0.2 per year.

The virtue of this statistic is that it is readily understood in terms of the fraction exposed who suffer the

Table 1. Individual risk of death vs age for UK, all causes (Grist, 1978)

Age group	Individual risk per year ($\times 10^{-3}$)
0–4	3.3
5–9	0.3
10–14	0.3
15–19	0.6
20–24	0.7
25–34	0.8
35–44	1.8
45–54	5.8
55–64	14.8
65–74	36.7
75–84	87.7
85 plus	205.2

specified effect. The problem arises in establishing an acceptable definition of the population at risk. Whilst we can readily agree that we are all about equally susceptible to certain kinds of disease, certain groups are known to be more susceptible than others to, say, heart disease or suicide.

Within a given category of risk, e.g. road accidents, we may identify a sub-group who are exposed to a higher level because they adopt a particularly hazardous form of transport, e.g. motor cycling. We need to distinguish between individual risk for the population at large, as calculated by Grist, and individual risk for an identified group who are recognisably preferentially exposed. Bearing in mind the above definition of individual risk we need to distinguish two meanings — one is the fraction of the UK population who are motor cyclists and who suffer a fatal accident and the other is the fraction of motor cyclists who get killed.

A further refinement of meaning arises if we make individual risk specific to certain occupations. For example, the individual risk of death amongst underground coal-miners is about 2×10^{-4} per year amongst that particular sub-group of the population. It is important to realise that in this sub-group each individual puts in about the same number of hours, so that all of the members are roughly equally exposed. This could not be the case for activities where the individual's annual participation varies enormously, as in motoring for example. Consideration of loss rates associated with activities for which the level of participation is readily quantified leads us to the next topic.

Fatal accident frequency rate (FAFR)

FAFR is so well known that little needs to be said concerning its use. Kletz (1971) has developed FAFR, which is defined as the number of fatalities suffered per

100 million man-hours of work at a given activity. It has proved to be of special value in assessing occupational risks, but can equally well be applied to other kinds of activity provided the number of man-hours of exposure can be determined.

In principle one could convert individual risk to a form of FAFR by calculating the number of man-hours of being alive enjoyed by a given total population over a one-year period and dividing that figure into the number of fatalities. However, this would simply be changing the units; the relative levels would be unchanged.

Forms related to FAFR

One can readily appreciate that FAFR gives a measure of risk of death per unit of risk-significant activity (i.e. man-hours of exposure); the unit of benefit to the risk-taker is also implicit in that he presumably works to earn money, and his wage increases with man-hours put in. However, there are some activities for which the unit of risk-significant activity and the unit of benefit are not so conveniently related. To illustrate this difficulty one may consider the various ways of expressing the risk of fatal accident whilst travelling. Sowby (1964) calculated deaths per 1000 million man-hours of travel for aircraft, bus, rail and motor car transport (a concept similar to that of FAFR) and compared the results with corresponding calculations of deaths per 1000 million man-miles of travel (Table 2). In terms of man-hours the high-risk category was aircraft, followed by motor cars with bus and rail well below these. On the man-miles basis the order was motor cars, followed by aircraft, with bus and rail again well down the scale. Now the question arises as to whether these are fair comparisons. Taking aircraft first, the majority of accidents are associated with taking-off and landing; the number of *fatal accidents per operation* (i.e. taking-off or landing) averages about 10^{-7} for the US carrier fleet (Rasmussen, 1975), and it would seem that this is the risk-significant

Table 2. Travel risks on per mile and per hour basis
(Sowby, 1964)

	Death rate	
	per 10^9 passenger hours	per 10^9 passenger miles
Air (ICAO states)	2400	11
Bus		
USA	80	1.7
GB	30	1.3
Rail		
USA	80	1.3
GB	50	1.3
Car		
USA	950	24
GB	570	19

activity. The apparent risk expressed in passenger-miles
or passenger-hours could be reduced simply by spending
more time cruising the aircraft on a circuitous route
between take-off and landing points, because this would
not significantly increase the number of accidents. However, if we consider the unit of benefit for the passenger,
the majority of people fly in order to go from A to B
quickly and cheaply. On that basis one could propose
that the risk in terms of *perceived benefit* should be
expressed as *deaths per passenger-mile-per-hour per unit
cost of fare* (or some weighted relationship between
these variables), whilst the risk in terms of *loss-significant
activity* should be *deaths per operation*.

The case would be different when applied to motoring; the perceived benefit could still be expressed in
terms of passenger-mile-per-hour per unit fare for those
simply anxious to get from A to B, but sightseers would
need a different calculation. The risk-significant activity
is different to that for aircraft, since here one would
expect to observe an increase in fatalities as more miles
are travelled and as the average speed of travel increases.

Thus the risk-significant variable is probably *vehicle miles per hour,* not *passenger miles.*

Clearly one needs to undertake a careful inspection of activities before one can quantify risk, and it seems that even for transport we cannot devise a scale that is broadly applicable. The distinction I wish to emphasise is between the unit of *perceived benefit* and the appropriate unit of *risk-significant activity.* Special care needs to be taken in defining the latter quantity, since a single form may not be commonly applicable to different classes.

Loss of life-expectancy

Several authors have used the concept of loss of life-expectancy as a means of expressing risk. Bowen (1976) considered the loss of life-expectancy for an individual risk of 10^{-5} per year and argued that this level of risk implied a rate of individual loss of life-expectancy that roughly balanced the observed rate of increase in life-expectancy with date of birth in the UK (due to improvements in social conditions and other factors). Martin & ApSimon (1974) considered the total number of years of life-expectancy lost in a population exposed to cancer-inducing radiation and calculated a cost in terms of an assumed value for a year of life for various discount rates.

The concept of loss of life-expectancy averaged over the population at risk has been developed by Hibbert (1978) and Reissland & Harries (1979). Hibbert obtained data on the age distribution of ex-coalworkers who died from pneumoconiosis during 1974 and calculated the total loss of life-expectancy suffered by those victims. He found that 449 individuals had collectively suffered 4871 years loss of life-expectancy. Reasoning that these deaths were incurred amongst a group of workers exposed over a period of years to conditions prevailing in the industry 20 to 30 years ago, Hibbert estimated the total group at risk to be 450 000. He then calculated the *average* loss of life-expectancy as 4871 years divided by 450 000 workers, which yields 3.95 days lost per worker exposed. Reissland & Harries have carried out similar calculations based on historical data for workers in various industries and have compared the results with estimates for workers in the nuclear industry.

This average loss of life-expectancy calculation is another way of treating raw information to yield a number, condensing the mass of input data to a statistic. The procedure is clear and well-defined, but I would argue that the resulting figure is highly misleading as a measure of real risk. By way of comparison the individual risk calculation has its shortcomings, but one can very readily extract the meaning. For example given the individual risk of death by lightning as 1 in 10 million per year in the UK then with 54 million inhabitants we readily interpret the average annual death toll as about 5, although we may have some uncertainty as to the number really at risk. None the less the meaning is clear. This cannot be said for average loss of life-expectancy. Referring to Hibbert's results for pneumo-coniosis, the apparent meaning, the one that a layman could be expected to infer, is that 450 000 workers each lost some quantity of life expectancy that can be fairly represented as 3.95 days on average, but not departing greatly from that figure in more than a few extreme cases. There is no sense in which this calculation can be said to represent the underlying fact that 449 specific victims lost 10.8 years each on average. One might call *this* calculation the average actual loss of life-expectancy, averaged over the identified victims. Similarly the *total* loss of life-expectancy (4871 years) is a telling statistic, but to express the risk in terms of 3.95 days lost on average expresses what is really a very substantial loss for each victim in such a way as to make it look negligible.

One can recognise here the same kind of problem that

is evident in the use of fN lines; 100 deaths in one incident occurring once every 10 years is less tolerable than a total of 100 deaths occurring one at a time spread out over the same period. In the case of life-shortening, one may well expect to find that 4871 man-years of life lost would be less readily accepted when distributed amongst 449 victims (10.8 years loss each on average) than when distributed amongst 450 000 (3.95 days loss each on average). The latter risk would probably be regarded as negligible, but the reality is expressed in the former figure. Given this likely non-reciprocity of acceptability one could not in fairness apply this calculation to compare risks without additional information.

Given the reservations I have expressed about calculations of average life-shortening as a measure of risk, it may appear that I do not perceive much value in the concept of loss of life-expectancy in risk appraisal. Later in this paper I propose a criterion for the risk of delayed death for incidents involving multiple casualties. The basis of this proposal is the use of total loss of life-expectancy as a risk quantifier; this seems to me to be a reasonable use of the concept in that it recognises that the loss for an individual may be years rather than days.

The frequency vs consequence line

It has now become common to express risk in the form of a line depicting the relationship between the frequency of occurrence and the magnitude of the consequences for a spectrum of events. The use of such lines as limiting criteria can be traced back to the work of Farmer (1967) who proposed a risk criterion for thermal nuclear reactors in this form where the consequence considered was the quantity of iodine—131 released. Beattie (1967) extended the interpretation by expressing the consequences of releases on this limit line in terms of the number of casualties.

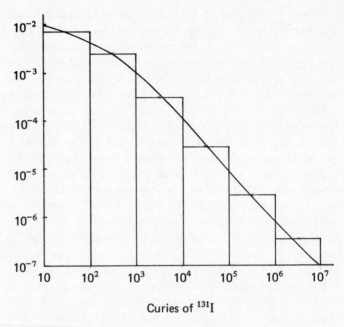

Curies of ^{131}I

Figure 1. Interpretation of the Farmer Curve.

Much discussion has been devoted to the interpretation of the Farmer type of fC line (Beattie, Bell & Edwards, 1969; Bell, 1970; Farmer, 1972, Meleis & Erdmann, 1972; Smith & Kastenberg, 1976) but the definitive interpretation given by Farmer and his co-workers is that the frequency associated with a particular value of consequence C represents the limiting frequency of occurrence for events with consequences roughly in the range $(C/3)$ to $(3C)$, or approximately a factor of 10 spanning the central value of C. It is in this same form that Kinchin (1978) originally expressed his proposed criteria for nuclear reactor accidents, with the consequence considered being the number of deaths, N. A further feature of Kinchin's criteria is that the product f times N is constant for a given line. It should be carefully noted that this form of fN line associates a frequency with a certain interval on the consequence axis, so that one may consider the Farmer-type fN line as one that joins the midpoints of the bars on a histogram (Figure 1). This distinguishes it from the cumulative form of fN line which shows the frequency of occurrence for events in which N or more deaths are suffered. The cumulative form has been used to express the results of *a priori* estimates of risk, e.g. the Rasmussen Report (1975) and the Canvey Report (Health and Safety Executive, 1978) and to collate historical data on the incidence of multiple-fatality accidents (Griffiths & Fryer, 1978; Fryer & Griffiths, 1979). Given a precise definition, one can readily convert from one form to the other, but there are certain advantages in the cumulative form. To begin with one may consider the expression of historical data on multiple-fatality accidents. The raw data consist of a list of events with dates and numbers of fatalities. One selects a data period of so many years and proceeds to calculate frequencies for events with N or more fatalities simply by adding up the number of events recording N or more fatalities and dividing by the data period. The first problem to arise is how to pick the values of N at which to calculate f. If one has information on events where N is accurately known, the 'obvious' answer is simply to calculate f at values of N dictated by the data. In practice one finds that this original accuracy is lost in the width of the line on the graph at large values of N on a log–log scale. The other major problem is that N is often not known accurately and different sources give different numbers for the same event. Fryer & Griffiths (1979) found it best to specify prescribed sorting intervals and to use these for all the categories of data. The cumulative fN line lends itself well to the representation of historical data because clearly one will encounter 'gaps' in the values of N, that is ranges of values of N for which no accidents have occurred. These ranges would have corresponding zero values of frequency on an interval form of fN line and the result would be a non-continuous line with occasional zero values. This form of representation is very cumbersome and does not aid the understanding of comparisons between different classes of event. Using the cumulative representation, one obtains a continuous line, since the absence of events in any range simply reduces the *slope* of the line to zero and one can visually compare the frequencies of occurrence in a given range by inspecting the slopes of the graphs.

Turning now to the viewpoint of a designer operating within the constraints of an fN line as a criterion, it is convenient to consider consequences in interval ranges because of the uncertainties in reliability and consequence modelling and because of the complexity of the system. Thus the designer favours the interval form of fN line. As already noted it is easy to convert from one form to another given a precise formulation of the original, and I would now like to propose that we adopt the cumulative form as a universal practice in expressing risk. The reason for choosing the cumulative form, apart from the minor factors of convenience in use referred to above, is that the cumulative form is intrinsically precise and has only one interpretation, whereas the Farmer-type interval form is open to two extreme interpretations, as shown below.

Suppose we take a Farmer-type fN line where the frequency $f(N)$ is to be interpreted as that associated with events giving fatalities in the range N_1 to N_2 where $N_2 = 10N_1$ and the range is logarithmically centred on the value N. Thus $N_1 = N/\sqrt{10}$ and $N_2 = \sqrt{10}N$. We will take the case where the product f times N equals a constant, i.e. $f(N) = k/N$, so that the line has a slope of -1 on a log–log scale. The argument applies just as well to other slopes, but this is the one most commonly used and it has the advantage of a minimum requirement of symbols to express the equations. Starting with the first decade interval ($N_1 = 1$, $N_2 = 10$) the midpoint value is $N = \sqrt{10}$ and the corresponding value of f is $k/\sqrt{10}$. For this segment of the line the total annual average death toll, t, is the product of frequency and the number of fatalities. However, the criterion permits this number to be anywhere in the range N_1 to N_2, so there are two extreme interpretations that can be given to t:

$$t_{max} = f(N)N_{max}$$
$$t_{min} = f(N)N_{min},$$

where $N_{max} = N_2$, $N_{min} = N_1$ in the first decade interval on the N-axis. These equations apply to all the decade intervals on the N-axis, but, noting that $f(N) = k/N$ and that $N_{max} = \sqrt{10}N$, substitution yields

$$t_{max} = (k/N)\sqrt{10}N = \sqrt{10}k;$$

and, since $N_{min} = N/\sqrt{10}$,

$$t_{min} = (k/N)(N/\sqrt{10}) = k/\sqrt{10}.$$

If we establish a value N_L as the largest value of N for which the criterion is intended, then adding the contributions from each decade interval we obtain the total annual average death toll, T, with maximum and minimum interpretations

$$T_{max} = \sqrt{10}k(\log_{10}N_L)$$
$$T_{min} = (1/\sqrt{10})k(\log_{10}N_L)$$

From these equations we see that

(*a*) the maximum and minimum interpretations differ by a factor of 10, which corresponds to the width of the interval specified in the criterion definition;
(*b*) the interpretations of T depend on the largest value of N for which the criterion is intended to apply;
(*c*) it is necessary to specify a particular interpretation of T (perhaps the logarithmic mean) before one can convert to an equivalent cumulative form of fN line.

Further difficulties of this kind arise if one assumes that the point at which f should be evaluated is, so to speak, allowed to slide along the line rather than be restricted to the logarithmic mid-points of each decade interval.

Turning now to the cumulative form of fN line as the starting point, one specifies a frequency $f_c(N)$ associated with the occurrence of events in which there are N or more fatalities. We will take the same form, i.e. a line of slope (-1) so that

$$f_c(N) = k_c/N,$$

where k_c is the constant for the cumulative case. It is easily shown that this can be converted to the interval form as a line having the same slope, but with a different constant, k, and yielding a unique interpretation for T.

The average frequency associated with values of N in the interval N_1 to N_2 (inclusive) is simply obtained by inserting N_1 and $(N_2 + 1)$ and subtracting, thus

$$f_{(N_1 \to N_2)} = k_c \left\{ (1/N_1) - [1/(N_2+1)] \right\}$$

The total annual average death toll too has an exact meaning. The frequency f_N associated with exactly N fatalities is given by the same expression with values N and $N + 1$ inserted,

$$f_N = k_c \left\{ (1/N) - [1/(N+1)] \right\}$$

or

$$f_N = k_c \left\{ 1/[N(N+1)] \right\}$$

The total annual average death toll is then simply the product of this frequency and the number of deaths N, summed over all values of N up to the largest value N_L, so that

$$T = k_c \sum_{N=1}^{N_L} [1/(N+1)] = k_c S$$

Table 3 shows values of the summation term S for various values of N_L (one decimal place would be more than sufficient in practice). There are numerous intricacies that arise from the lack of precise definition

Table 3. Values of the summation
term S for various values of N_L

N_L	S
10	2.019
10^2	4.197
10^3	6.486
10^4	8.787
10^5	11.090
10^6	13.393

referred to here. For example one can establish a relationship between k and k_c based on the requirement that the average frequencies of occurrence associated with a given interval be equal in the two forms of line. This depends on the slope of the line and the size of the interval specified. One may calculate the unique value of T derived from the cumulative expression and then examine the coefficient required to obtain the same value from the Farmer-line expressions for T. One finds that this is *not* obtained by assuming $T = k(\log_{10} N_L)$ as one might assume; it is necessary to take a value closer to three-quarters of this expression in order to obtain equality, and the factor required is sensitive to N_L. All of this complication can be avoided if one uses fN lines in the cumulative form from the outset.

There must be few discussions on risk that do not make reference to multiple-fatality accidents as a topic of special concern. The problem has often been encapsulated in statements to the effect that society is prepared to tolerate a heavy annual death toll if the incidents involve only one or two deaths at a time, but special concern surrounds events in which many deaths are suffered simultaneously. Looked at from the point of view of the average death rate the accident that takes 100 lives once every 10 years is the same as 100 single deaths spread over 10 years, but the former case appears to generate much greater concern about the risks involved. This problem raises the question of how life is valued. Wilson (1975) places a value on this concern and assumes that 'a risk involving N people simultaneously is N^2 (not N) times as important as an accident involving one person.' There is no compelling reason to accept this particular arbitrarily chosen weighting, but one can point to reasons why simultaneous fatalities should be regarded with greater concern, e.g.

(a) If the deaths occur at one time then the immediate social impact on families involves many people simultaneously, thus making the scale of the loss appear more significant against the steadier background of smaller events.

(b) If the victims are all drawn from one community then the impact is less readily absorbed than if many separate groups were involved.

(c) If the deaths occur all in one place then the scale of the loss is more apparent than otherwise (it is conceivable that a single incident could lead to many delayed deaths in diverse locations).

(d) If the deaths are all attributable to one event then there will be understandable concern that one mistake or one fault should have such large consequences. However, it is notable that this may not necessarily apply if the deaths are attributable only to one *kind* of activity, e.g. consider the apparent acceptability of 7000 deaths a year in the UK from motoring accidents.

On the basis of such considerations one may argue that diversity of geographical or community origins of the victims would increase the acceptability of a given multiple-fatality event. For example, most aircraft accidents would involve people from many diverse groups, whereas a dam failure would involve a more localised group of communities. Whatever the outcome of such considerations may be, it seems that there is an emerging recognition that some weighting needs to be introduced in risk criteria to express this non-reciprocity between frequency and magnitude of consequences and that a line of equal *fN* product may not be an appropriate form of criterion for early deaths.

Proposal for the relationship between fN criteria for early and delayed deaths

One of the outstanding developments in hazard awareness over the last two decades has been the increasing recognition of the importance of delayed effects in the assessment of potential damage. This trend has been well reflected in the emergence of regulatory legislation (especially in the USA) specifying standards and practices for the handling of numerous materials that are believed to have carcinogenic, teratogenic or mutagenic effects. The objectives of industrial safety regulation have thereby been expanded beyond the traditional concern to avoid immediate injury within the factory boundary; in the extreme we must now consider potential damage from agents that may not manifest their effects until well beyond the lifetime of the present generation and even beyond the expected lifetime of current civilisations and their institutions.

A restricted but significant facet of this long-term aspect of risk concerns the development of a criterion for effects that may lead to the delayed death of an individual. Such effects can result from exposure to ionising radiation, producing latent cancers, as well as from exposure to certain chemicals. The effects of chemicals in this context are much less well-understood than those of radiation.

In attempting to propose a criterion for the risk of delayed death one must be sensitive to the possibility that the public may be indifferent to the latency of a death risk if its cause can be traced to a particular industrial activity. A criterion based on such a view would then be the same as that for immediate deaths. However, few people would argue that death at some time in the future is preferable to immediate death, and criteria have been proposed by Kinchin (1978) in the form of separate, Farmer-type, fN lines for early and delayed deaths, the frequency of occurrence proposed for the delayed-death criterion being a factor of 30 greater than that for early deaths.

The criterion for delayed deaths proposed here is expressed in the form of a frequency, f, vs number of deaths, N, line but differs from Kinchin's both in its basis and in its result. Reference has already been made to the concept of loss of life-expectancy and its various uses in risk assessment. The argument presented here makes further use of this concept and proceeds as below.

Suppose that an individual is aged i years (i.e. he has completed i years) and that he may expect to live to age $D(i)$. Then his remaining life-expectancy, $E(i)$, is $(D(i) - i)$ years. If we now suppose that he suffers exposure to some agent that will bring about his death at the end of a total period of $I(i)$ years his age at death will then be $(i + I(i))$ years, provided that $I(i)$ did not exceed his unmodified remaining life-expectancy $E(i)$. The loss of life-expectancy, $L(i)$, is

$$L(i) = D(i) - [i + I(i)]$$

and if there are $N(i)$ individuals aged i who suffer this effect then the total loss of life-expectancy L_T in the exposed population is

$$L_T = \sum_{i=0}^{D(i)-I(i)} N(i)[D(i)-i-I(i)] .$$

The upper limit on i is fixed by the condition that loss of life-expectancy is suffered only by those individuals for whom the unmodified remaining life-expectancy $(D(i) - i)$ is greater than or equal to the new period to termination $I(i)$, that is

$$D(i)-i \geqslant I(i)$$

or $$i \leqslant D(i) - I(i)$$

whence $$i_{max} = D(i) - I(i).$$

In order to evaluate L_T in a rigorous fashion we would need to insert expressions specifying the age distribution of the population, $N(i)$, the distribution of values of un-modified expected age at death, $D(i)$, for individuals currently aged i years and the distribution of values of $I(i)$, the latency period, which additionally would itself vary with i. This could be done, although it might be difficult to be confident that $I(i)$ had been accurately estimated. However, it is probably sufficient to use average values and to assume a uniform age distribution. Following this simple approach we suppose that all individuals would normally die at age D (i.e. $D(i) = D$) and that the age distribution of the population is uniform so that if the total number of individuals exposed is N_T then the number aged i, $N(i)$, is simply (N_T/D) for all values of i. Further we suppose that $I(i)$ can be replaced by an average value I. The loss of life-expectancy then reduces to

$$L_T = \sum_{i=0}^{D-I} (N_T/D)(D-I-i).$$

Under these assumptions the term (N_T/D) can be taken outside of the summation; the bracketed-term summation is simply the sum of the integers from 1 to $(D - I)$, for which there is a standard result $(S_n = n(n + 1)/2)$ which yields

$$L_T = N_T (D-I)(D-I+1)/2D.$$

We now state the basis of the criterion proposed here which is that the limiting frequency of occurrence for delayed deaths, f_D, should be such that the *total annual average loss of life-expectancy* is the same as for incidents resulting in early death, for which the frequency of occurrence criterion has been specified as f_E. For early deaths we have $I = 0$, so that the proposal may be expressed as

$$f_E N_T D(D+1)/2D = f_D N_T (D-I)(D-I+1)/2D,$$

from which we obtain the ratio R

$$R = (f_D/f_E) = D(D+1)/(D-I)(D-I+1).$$

In determining the values of I to be used in this expression one may refer for example to the induction of cancer following irradiation. A simplified description is that following such exposure there is a period during which there is no increase in the incidence of cancer in the exposed population but this is followed by a period of increased cancer incidence that lasts for a number of years. The incidence period varies with the type of cancer. For leukemia it is probably from 5 to 20 years after irradiation, whilst for other forms 10 to 40 years is more likely. As previously noted, these distributions of I, which are also functions of age i, should enter into a rigorous evaluation of the expression for loss of life expectancy, but here we will use average values. Table 4 lists values of the ratio R evaluated with $D = 70$ years for average values of I ranging from 5 to 60 years. The choice of a criterion value for the ratio R clearly depends on specifying a representative value for I, but we may note that even for $I = 30$ years, which is towards the end of the likely incidence period for radiation induced cancers other than leukemia, the frequency, f_D, for delayed death incidents is only three times that for early deaths.

It may be argued that the criterion proposed above is deficient in that it takes no account of the fact that there will be little explicit impact during the early years of an industry and that it may take many times the average induction period before a 'steady-state' impact starts to manifest itself. This is a fair criticism since one can envisage that the significant lifetime of a particular industry might be quite short — for example, parts of the energy industry might be replaced within 50 years as new methods gain commercial viability. However, one cannot predict such developments with any accuracy and it seems reasonable to treat all cases on the same basis, which is to assume that a steady state does exist.

In proposing this relationship between early and delayed death criteria it is necessary to give some attention to the question raised previously concerning whether or not a line of constant fN is likely to be a line of constant acceptability. For early deaths it seems unlikely that this can be so, since it is reasonable to argue that the criterion ought to be weighted against larger values of N to allow for the greater social impact suffered if those N deaths occur in one community, in one location, at one time and due to one cause. Referring again to motoring deaths we note that, even though there is a common cause, 7000 deaths a year are tolerated in the UK from this single cause, but they are spread over many communities, locations and times. For delayed deaths for a single primary incident it is clear that the passage of time that ensues before death occurs will lead to considerable dispersion of the social impact, firstly in

Table 4. The ratio $R = f_D/f_E$
for various values of I

I (years)	R
5	1.1
10	1.4
15	1.6
20	1.9
25	2.4
30	3.0
35	3.9
40	5.3
45	7.6
50	11.8
55	20.7
60	45.2

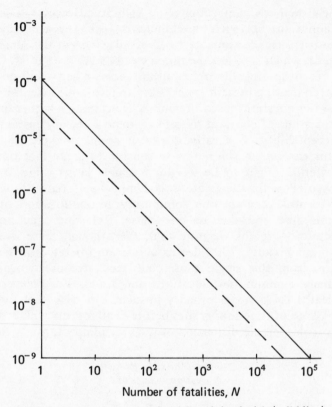

Figure 2. Proposed criterion for delayed death risk (solid line) compared with Kinchin's proposed criterion for early deaths (broken line) for one reactor. Both are expressed in cumulative form.

the timing, secondly in the location and thirdly in the community affected. The degree to which these dispersions dilute the impact will depend much on the degree of mobility within the community, but it is arguable that a line of equal fN is a fair criterion for delayed deaths. Finally, to give substance to this proposal I have expressed it in terms related to Kinchin's criterion for early deaths for a single nuclear reactor. In Figure 2 the Kinchin line for early deaths is shown for a single reactor. I have chosen I = 30 years as a good compromise, and thus my delayed-death criterion line for a single reactor appears as a line of slope – 1 on a log–log scale with the product $fN = k$ where $k = 10^{-4}$.

References

Beattie, J. R. (1967). Risks to the population and the individual from iodine releases. *Nuclear Safety*, 8 (6), pp. 573–6.

Beattie, J. R., Bell, G. D. & Edwards, J. E. (1969). Methods for the evaluation of risk. *UKAEA Report AHSB(S)R159*. Her Majesty's Stationery Office, London.

Bell, G. D. (1970). Safety criteria. *Nuclear Engineering and Design*, 13, Chapter 2.

Bowen, J. H. (1976). Individual risk vs public risk criteria. *Chemical Engineering Progress*, February, pp. 63–7.

Farmer, F. R. (1967). Siting criteria – a new approach. *Atom*, (128), pp. 152–70.

Farmer, F. R. (1972). Letter to the Editor. *Nuclear Safety*, 13 (5), pp. 362–4.

Fryer, L. S. & Griffiths, R. F. (1979). World-wide data on the incidence of multiple-fatality accidents. *UKAEA Report SRD R149*. Her Majesty's Stationery Office, London.

Griffiths, R. F. & Fryer, L. S. (1978). The incidence of multiple-fatality accidents in the UK. *UKAEA Report SRD R110*. Her Majesty's Stationery Office, London.

Grist, D. R. (1978). Individual risk – a compilation of recent British data. *UKAEA Report SRD R125*. Her Majesty's Stationery Office, London.

Hibbert, J. (1978). The incidence of mortality expressed in terms of life shortening. *NPC Report, NPC (R) 1304*.

Health and Safety Executive (1978). *Canvey – an investigation of potential hazards from operations in the Canvey Island/ Thurrock area*. Her Majesty's Stationery Office, London.

Kinchin, G. H. (1978). Assessment of hazards in engineering work. *Proceedings of the Institute of Civil Engineers*, 64, pp. 431–8.

Kletz, T. A. (1971). Hazard analysis – a quantitative approach to safety. In *Major loss prevention in the process industries*. Institute of Chemical Engineers, Newcastle-upon-Tyne.

Martin, A. & ApSimon, H. (1974). Population exposure and the interpretation of its significance. *IAEA SM 184/9*, pp. 15–26.

Meleis, M. & Erdmann, R. C. (1972). The development of reactor siting criteria based upon risk probability. *Nuclear Safety*, 13, pp. 22—8.

Rasmussen, N. (1975). *Reactor safety study*. USNRC, Washington D.C.

Reissland, J. & Harries, V. (1979). A scale for measuring risks. *New Scientist*, 72, pp. 809—11.

Smith, C. F. & Kastenberg, W. E. (1976). On risk assessment of high level radioactive waste disposal. *Nuclear Engineering and Design*, 39, pp. 293—333.

Sowby, F. D. (1964). Radiation and other risks. In *Proceedings of the symposium on the transport of radioactive materials, Bournemouth, April 1964.* (Also *UKAEA Report AHSB (A) R8.*)

Wilson, R. (1975). The costs of safety. *New Scientist*, 68, pp. 274—5.

5 Dealing with hazard and risk in planning [1]

B. J. Payne

Scope of paper

Let me say at the outset that I make no pretence at being an expert on hazard or hazard risk. Instead, I welcome the opportunity of making a few comments from the standpoint of a town planner who is involved in the preparation of local plans in which the question of hazard potential is a fairly central issue; local plans concerned with the expansion of the Stanlow oil and petrochemical complex and its relationship to nearby residential areas.

Although this work is still under way I feel there are some useful comments to be made from our experience so far, and comments also arise from the countywide survey of chemical and allied industries carried out on our behalf by Cremer and Warner in 1975. These comments will I hope tell you something about the information and advice needed by planning authorities on hazard matters, if they are to contribute effectively to the reduction of risk to the public or the minimisation of possible hazard consequences.

My contribution will be mainly concerned with the

forward planning aspects of planning authority work. Chris Brough from Halton Borough will be dealing with development control and the questions arising from that, since it is the district planning authorities who have the lion's share of the power to determine planning applications.

Both our sets of comments are offered in a constructive rather than derogatory spirit. One of the reasons for holding this conference is a belief that present systems need improvement, so we all need to know what we really think of the present arrangements and where they need improvement.

Broad functions of town and country planning

Before going on to describe some of the work that we are involved in in Cheshire, I will briefly outline my views about the broad functions of planning and its special characteristics.

It aims among other things to achieve a satisfactory ordering of land use which minimises the degree of conflict between incompatible uses and sets aside land for likely future requirements of various kinds. This is achieved through the mechanism of development plans and development control. In the development plan system the authorities actually have the responsibility of planning ahead and establishing appropriate policies, while the development control system is basically a means of ensuring that public and private development takes place generally in accordance with the policies established in the development plan. The development plan comprises the structure plan prepared by the county planning authority and more detailed local plans for particular areas or topics, the great majority being prepared by district councils.

Planning is concerned among other things to improve levels of 'amenity', a word which covers a number of

environmental matters. Plans try to minimise the impact
of a variety of adverse factors like noise, smell, air pollution, visual intrusion and hazard by, for example,
making the necessary land allocations well away from
other sensitive uses. The required degree of separation
depends on a number of factors including public expectations about their local environment and the degree
of success achieved by the statutory authorities in
controlling emissions and noise levels etc. from particular
processes.

Planning obviously has to operate in the real world,
so the degree of separation actually achieved between
'incompatible' uses depends on a number of additional
factors such as the historical development of the area
(e.g. conditions may have to be accepted near an old-
established complex which would not be tolerated in
the case of a 'greenfield' development) or the degree to
which the area desires the new investment and jobs (this
obviously has an important influence on the bargaining
power of the authority).

The planning system provides the means for saying
'go' or 'no go' to particular types of development in
particular locations. The operational control on the
actual development is then generally with other authorities, for example the Health and Safety Executive or
the Environmental Health Authorities. This seems a
clear distinction, but in practice it must be recognised
that there will be areas of overlapping interest where
potential for misunderstanding, mistrust, etc, between
the different authorities exists and so the need for
adequate liaison and consultation must be fully appreciated by the different authorities.

Special characteristics of the planning system

In the belief that it is sometimes necessary to state the
obvious, the first point I make here is that *planning*

authorities must plan. In other words planning authorities cannot simply sit back and respond to proposals
put to them; at the same time they have the obligation
to plan ahead. They must look forward over a period of
about 15 years, formulate appropriate policies and proposals for land allocations, environmental improvements
etc. In doing so they make certain pronouncements
about their aims, about the conditions they regard as
satisfactory or unsatisfactory, and what, if anything,
they are proposing to do to better the situation.

This requirement to plan ahead means, for example,
that the authorities face the task of earmarking areas for
future development by heavy or potentially hazardous
industry before the details of the actual development
are known, so it may be necessary for the planning
authority to make some 'worst case' assumptions about
the impact of that development before deciding whether
or not to allocate the land for development.

My second point refers to *public participation.* A
significant requirement of the forward planning system
is that the general public must be given the opportunity
to contribute to the formative stages of plan making.
This in my view has a particular importance since it has
created certain expectations about the extent of public
involvement in other public decision-making spheres too,
but in any event it has established a climate in planning
departments under which most documents are prepared
with eventual publication in mind.

The development control system also is increasingly
open to public scrutiny and the situation in which the
planning officer can make his recommendations in
confidence has more or less disappeared.

The requirement of public participation means that
difficult and controversial issues will have to be thrashed
out in public. This can present tactical problems, such as
the way to consider alternative lines for a new road
without causing blight along all of them, or the way to
avoid unnecessary public alarm when the topic of major

hazard installations is being considered. The question of blight, or at least a reduction in property values, as a result of public discussion about levels of hazard risk being experienced in particular residential areas is clearly a very difficult problem and could all too easily lead to complete inaction even where levels of risk are privately regarded as unsatisfactory.

Work in Cheshire

Having set the scene in very general terms I now move on to describe the various pieces of work which have been going on in Cheshire since the Flixborough accident which are either directly concerned with hazardous uses or in which the question of hazard plays a major part.

The various elements are shown in Figure 1 (which is in fact simpler than at first appears!). Working down the page, the activities on the left-hand side are those taking place at national level, and on the remainder of the page those activities taking place in Cheshire.

The Flixborough accident is used as a salutary starting point (although Halton Borough had already thought about engaging Cremer and Warner on a consultancy basis to give technical advice on planning applications prior to the accident).

Immediately after the Flixborough accident Cheshire County and District authorities met to decide what action might be needed, being uneasily aware that some of the many chemical installations in Cheshire might be equally prone to disaster. It was then that Cremer and Warner were retained to give advice as required on planning applications and employed also to carry out a general background survey of the chemical and allied industries including an appraisal of their hazard and pollution potential. The County and District authorities felt that it was important for them to have a general understanding of the industry within their area and the

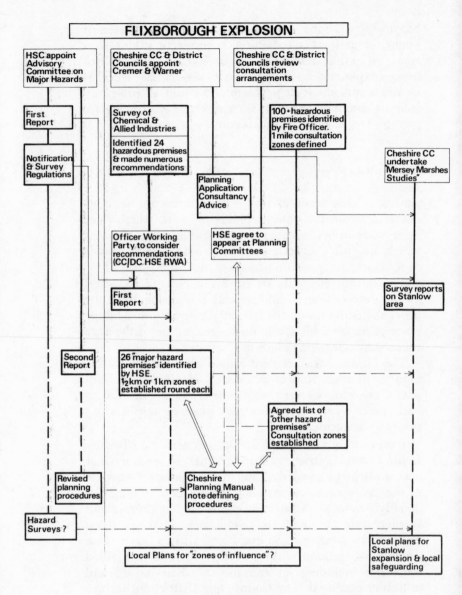

Figure 1. Cheshire's response to the Flixborough disaster.

necessary information was not available from the Health and Safety Executive, or elsewhere within the government sphere. Likewise with planning applications, the authorities needed appraisals of significant planning applications to help them judge the merits and problems involved but the amount of information available from the statutory authorities seemed inadequate for the purpose.

At the same time, County and Districts set about reviewing their consultation arrangements for planning applications involving or near hazardous premises, and a list of over one hundred such premises was drawn up in conjunction with the County Fire Officer. This list was treated as confidential, but each district Chief Planning Officer was notified of the premises in his area and was given a set of maps indicating a notional 1 mile consultation zone around each. This list is now being superseded by a refined list which we are drawing up at the moment with the help of the Health and Safety Executive.

The Cremer and Warner Survey produced a list of 24 major hazard premises which were mentioned by name in the published main report. The report contained a large number of recommendations for action by the authorities including one that a 'corporate action plan' be prepared. A working party was then set up to consider all the recommendations in detail, and this included representatives from the Health and Safety Executive in addition to the county and district authorities and water authorities. This work is still continuing and is taking on board the conclusions of the HSC's Advisory Committee on Major Hazards. We are currently trying to agree consultation zones of ½ km and 1 km around 26 premises which have been identified by the Area Director of the Health and Safety Executive as Major Hazard Installations within the terms of the draft Notification and Survey Regulations. That this number should nearly correspond with the number of premises identified in the published volume of the Cremer and Warner report is coincidental, since Cremer and Warner were mainly concerned with the chemical industry so the lists are not the same. We understand that the number of premises defined could rise to between 30 and 40 as a result of the notification received under the new regulations. This will leave an as yet undefined number of premises which do not constitute 'major' hazards yet whose existence and hazard potential ought to be taken into account when planning decisions affecting the immediate area are being taken.

When the list of and consultation zones for hazardous premises have been agreed we are intending to prepare a Practice Note for use by planning staff indicating the status of the sites (e.g. major hazard identified under the new regulations, or a 'minor' hazard identified locally), the availability of information about them, legislative arrangements, the procedures to be adopted over planning applications or when a local plan is being prepared, the need to consult the Health and Safety Executive, Fire Officer, Environmental Health Officer, etc. The Note will aim to ensure that planning staff are at least alerted to the existence of hazardous premises in their area so they can handle planning matters accordingly and can then go on to consider whether positive action might be needed in any of the affected areas.

It is worth mentioning that the employment by the County and District councils of outside consultants gave rise to some unease among some of the firms in the county, and the County Council received letters to that effect. The particular firms felt that a survey by technical consultants could cause breaches of commercial secrecy and that the authorities should rely on the services of government agencies. The survey went ahead as planned because the type of information being sought was not available elsewhere. However, Cremer and Warner did not receive 100% co-operation and in several cases had to enter into confidentiality undertakings. Subsequent discussions with the Health and Safety

Executive arising from their representation on the working party which was formed to consider the Cremer and Warner recommendations led to the Alkali Inspectorate and Factory Inspectorate agreeing to appear in public at Planning Committees when requested to supplement their advice on planning applications. In these circumstances it was decided that the joint retainer arrangement with Cremer and Warner could cease and that individual districts would employ them as and when the need arose.

The information available in the Cremer and Warner report and through the working party is being applied to the *Mersey Marshes studies* which are concerned with the future expansion needs of Stanlow. In these studies large tracts of land north, east and south of the present Stanlow complex are under consideration and the effect of further development of these areas upon the villages of Ince, Elton and Thornton-le-Moors is being given close consideration. No consultant advice has been sought on the question of hazard risk, although it has been for certain other aspects of the studies. As can be seen from Figure 1, these studies are at the stage where survey reports have been prepared and we await with interest the reaction of the public when these reports are published. As far as hazard is concerned, the survey reports name the installations which fall within the major hazard criteria and a brief general description is given in each case of the hazardous plant, although no inventories of materials are quoted. The level of detail is limited and so this particular section of the report could well be criticised as inadequate when compared, for example, with the amount of detail made available to the public in the Canvey Report. The Canvey Report will in my view come to represent a significant benchmark in the matter of availability of information on hazardous installations.

An important aspect of the studies is that concerned with buffer zones or safeguarding zones. If further development takes place on the sites desired by the existing companies then safeguarding for two villages in particular will be needed. Our researches have shown that although no recognised separation distances are set down in the UK they have been elsewhere. For example, certain regional administrations in Germany and the Netherlands have recommended a separation distance of 2 km in the case of heavy industry, including oil refineries and chemical works. However, we have been informed by our Area Director of the Health and Safety Executive that if formal safety zones were established in this country (as distinct from consultation zones) they would be no greater than 200–300 metres round any particular plant. This surprised us, but our detailed work on the other environmental elements like noise, air pollution and visual intrusion was in any case leading us to the view that much greater separation distances would be needed if the general amenity of the residential areas is to be safeguarded. It is possible therefore that 'safety zones' could form a part of rather broader 'environmental safeguarding zones'. We are trying to determine these at the moment but it is already apparent that there is a wide divergence of view between ourselves and industry over the desirable degree of separation.

Conclusions: the notion of 'acceptability'

Most of what I have said in this paper is concerned not so much with the acceptability of risk as with the task in which we have been involved in Cheshire: ensuring that adequate background information exists about the industry that gives rise to many of the hazard risks, the chemical industry, and ensuring that foolproof procedures are adopted for alerting planners to the existence of hazards within their area from this and other industries and, therefore, alerting them to the need to obtain outside advice when planning proposals are under consideration for that site or in the vicinity. This sounds a modest task, but we are still not there 5½ years after Flixborough. Once this has been established, we are then very much in the hands of the experts for helping us to advise our elected members on what constitutes an 'acceptable risk'.

The Advisory Committee on Major Hazards put forward the view (with which we cannot disagree!) that the siting of all industrial developments should remain a matter for planning authorities to determine, since the safety implications, however important, cannot be divorced from other planning considerations. Elected members are thus asked to strike some kind of balance between the risks arising from a new development and its other merits. There clearly needs to be confidence that the advice they receive on risk is in fact expert advice, confidence not only among elected members but the local public as well, and confidence that the statutory bodies have adequate control over the ensuing development.

From my point of view I would hope on the one hand that the Health and Safety Executive can look ahead to the longer-term issues, for example discussing with the district and county authorities involved the planning implications arising from the hazard survey assessments carried out under the new regulations, and offering guidance on the general principles to be adopted in particular local plans where the future location of potentially hazardous activities is being considered.

I would hope on the other hand that observations made on planning applications can be reasonably meaty and informative, containing quantified and comparative information wherever possible. Chris Brough will have more to say on this particular point, so the only comment I would wish to add here relates to staff resources. Once the consultation zones and procedures are properly set up can the statutory bodies make the necessary staff time available to make their responses in sufficient detail and in the time required?

Whilst recognising the need for caution and discretion in handling information on hazard, I would make a plea for discussions to be held openly wherever possible, and for items like lists of hazardous installations and their consultation zones to be published information, for only in this way can it be guaranteed that proper consultation procedures will be installed and observed and the available information brought to the attention of those who need it. The debate really should focus on appropriate emergency procedures, appropriate remedial action, and appropriate future planning policies rather than whether or not the local residents are entitled to know that a particular hazard exists; this should be taken for granted.

Dealing with hazard and risk in planning [2]

C. W. Brough

Introduction

Bernard Payne has already set much of the scene for my part of our joint chapter. I can, therefore, be brief in introducing the subject matter of my presentation to you.

My aim today is to identify what I consider to be our main problem when planning applications involving major hazards are dealt with by planning authorities.

I hope to achieve my aim by first telling you a little about Halton and then quoting a few examples to you of planning applications involving major hazards which my council has dealt with. I will then draw out my conclusions. In my conclusions I do not deal with the question of legislative change. That is because of shortage of space and because I think legislative change is relatively simple to achieve. I have, however, produced a short appendix to this chapter suggesting some possible changes.

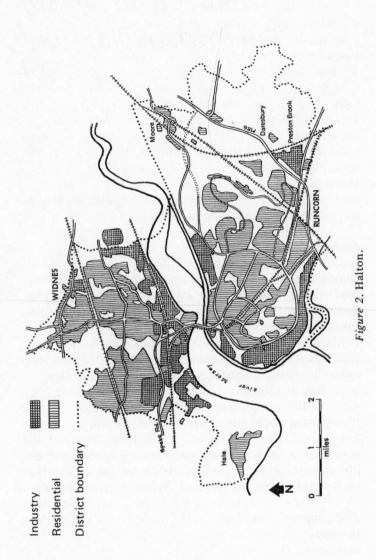

Industry

Residential

District boundary

Figure 2. Halton.

Halton

The Halton area
The name 'Halton' is local government's new name, invented in April 1974, for the towns of Runcorn and Widnes together with four surrounding rural parishes. The district straddles the Mersey estuary and is physically linked by the Runcorn—Widnes bridge (see Figure 2). The two towns are industrially linked through their equal and heavy dependence upon the modern general chemicals industry (Figure 3). This heavy dependence obscures the complexity and variety of life in the district but does partly explain our experience in dealing with complicated planning applications submitted by the chemical industry.

Halton Borough Council
Halton Borough Council comprises 47 councillors elected to run its functions on behalf of the 120 000 inhabitants of the borough. Business is divided between various committees. One of these committees deals, among other things, with applications for planning permission. It has fourteen members, four of whom have had direct experience of working for the chemical industry. It normally meets once every six weeks and decisions are invariably made in public (Figure 4).

Planning applications in Halton
Every year the council receives 900—1000 planning applications. About forty or fifty involve significant pollution or hazard issues. One or two each year have been for proposals for new or additional major hazards.

The Council's consultants
In April 1974 Halton Council formally came into existence. Officers and councillors quickly decided they could not adequately and competently determine some of the complicated large-scale industrial applications. A

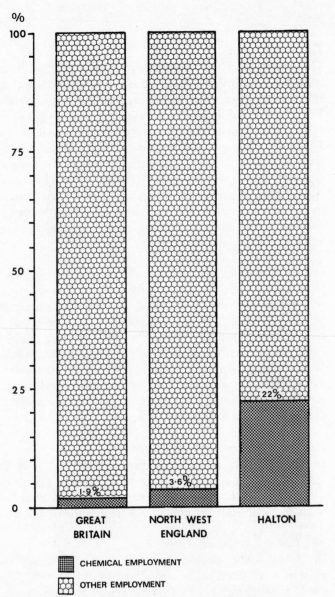

Figure 3. Employment in the chemical industry, 1976.

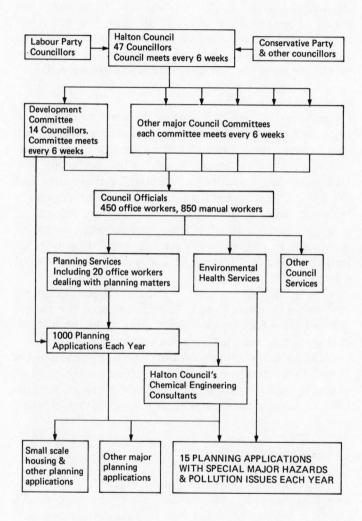

Figure 4. Halton Council: structure relating to planning applications involving major hazards.

decision was taken to employ Messrs Cremer and Warner as the council's chemical engineering consultants. They are still employed by the council and the arrangement has been reviewed annually. In the last five years our consultants have reported on an average of about fifteen applications a year. Eight of these over the whole five years have been for new or additional major hazards.

Major hazards in Halton

I mentioned earlier that one or two planning applications each year have been for major hazards. Many more applications have been for non-hazardous proposals within or close to 'major hazard' sites. In April 1974 Halton already contained a number of these sites and the eight major hazard planning applications since then have not radically altered the distribution of these sites. Using a 2 kilometre radius from each site as a convenient shorthand to express the potential area of concern we find that over 50 000 people live within 2 kilometres of a major hazard (Figure 5).

This situation, in part, explains Halton Council's concern at ensuring proper consideration of the hazard and risk elements in new developments in the district. It does not in my view explain wholly why the council decided to spend money on hiring private consultants. I would like, therefore, to turn to some case studies of planning applications in Halton for new major hazards since these cases illustrate why we employed consultants.

Halton's hazard cases

The multi-million pound proposal (Figure 6)

The first example I would like to explain concerned a major chemical company who discussed its proposal with council officers early in 1976. It was for a multi-million pound new production facility involving the use of materials which would constitute a major hazard.

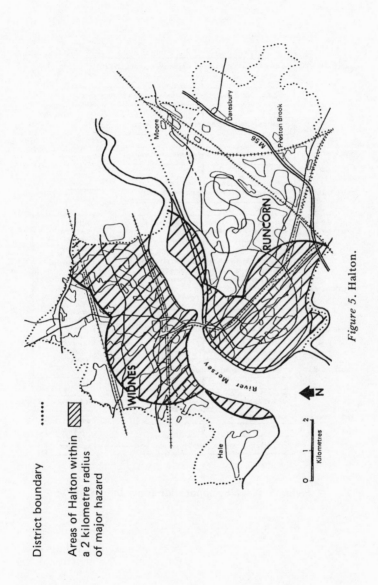

District boundary ......

Areas of Halton within a 2 kilometre radius of major hazard

Figure 5. Halton.

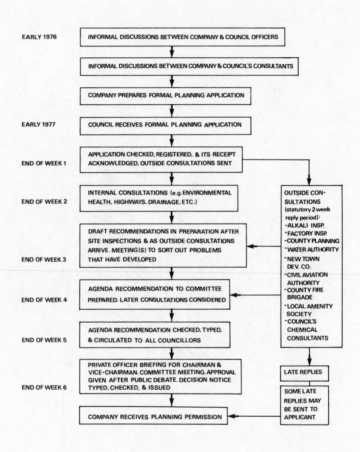

Figure 6. Planning application procedure example.

At this preliminary stage council officers advised the company how to go about making the best type of planning application administratively, and advised upon the various other public bodies that ought to be consulted. The council's chemical consultants started discussions soon afterwards with the company, at the request of council officers.

As a result of this preliminary contact, when the company's planning application was submitted, about a year later, approval was given in only six weeks. This decision was issued in six weeks even though consultations were made with six statutory bodies, including the Factory Inspectorate and also the local amenity society as well as the council's chemical consultants. I emphasise this speed because industrialists expect it, and normally get it, if they involve the council early on. I also emphasise speed because government ministers constantly ask for speedy decision-making from local planning authorities.

The proposal involved an investment of £25m and involved the storage of over 1000 tons of various hazardous materials. The only response received from the Factory Inspectorate was a preliminary and informal note. This note indicated that the Major Hazards Group had received preliminary notice about the process and had only made minor comments so far. The view was expressed that there would probably be no major objection to the proposals but work was still in hand. That information represented the only central government advice which my council received upon this matter. Fortunately, the council had the benefit of a report from its own consultants which dealt at some length with the various hazard issues involved and indicated a clear technical recommendation upon the matter. Without that advice I think my council would have been very reluctant to reach such a quick decision upon such a large scale, complex and potentially hazardous development.

Increased liquid chlorine storage

The second proposal related to a planning application received early last year. It was for increased production at another existing major hazard site. The increased production was, in itself, relatively harmless but it included a proposal to significantly increase the bulk storage of liquid chlorine. As before, the proposal had been informally discussed before submission. Similar consultations to the previous example were carried out, including the Factory Inspectorate and the council's chemical engineering consultants.

A reply was received from the Factory Inspectorate three weeks after being consulted. It stated that the papers had been submitted to the Major Hazards Group. They had replied that, in their opinion, the proposed development 'does not affect the existing nor create significant additional risks to persons on the premises or outside the curtilage of the site. There is, therefore, no objection made to this proposal'.

At roughly the same time the council received a detailed five-page report from its consultants. They expressed a clear view that they would prefer to see the expanded bulk chlorine storage relocated elsewhere within the site. The existing location of the tanks was very constricted, emergency accessibility was difficult and the restricted location, in their view, increased the risk of damage to the chlorine tanks and to associated pipe work. In view of the location of housing 400 metres away, a school 600 metres away, and a hospital 1 kilometre away, combined with the difficult plant location, the consultants expressed their technical view that the bulk chlorine storage facility would best be relocated. They also indicated that if the company carried out a full hazard and operability study in relation to bulk liquid chlorine storage then their technical reservations might be overcome. I think this difference in expert opinion may be related to the Health and Safety Executive considering that the state of the existing liquid chlorine storage was not a part of the new proposal. A planning authority on the other hand must always look at an existing situation integrally linked to a new proposal, as part of that new proposal.

My council resolved to approve the proposal, in principle, subject to the company submitting a full hazard and operability study to the council and to the company agreeing to incorporate any of the study recommendations into their proposal. The practical effect would be either to incorporate many sophisticated remote-control safety devices on the existing storage location or alternatively a complete move within the plant area to a more open location. The company informally agreed with the validity of the hazard issues raised but decided that it would be more economical to increase production without increasing bulk liquid chlorine storage. They therefore subsequently submitted a new planning application for increased production facilities without increased risk to life through additional, constricted, chlorine storage. This proposal was granted planning permission unconditionally.

Housing redevelopment

My third case relates to a proposal by my council to re-develop a nineteenth-century terraced housing area for new council housing. It was close to certain major hazards so, in August 1974, the council consulted the Factory Inspectorate under paragraph 15 of DOE circular 1/72. We received a detailed and sensible reply within one week! I quote from it:

In each of the first three cases the maximum credible incident would be the total failure of a single tank, together with, in the first case, the failure of a road tanker. There is little doubt that if an incident of this type were to occur in the first premises with a west to south west wind blowing there would be a considerable risk. . . In all these cases the probability of the maximum credible incident occurring is remote . . . objection to its redevelopment for residential use could not be sustained solely because of the risks associated with these installations.

This 'maximum credible incident' related to 27 tons of liquid chlorine about half a kilometre away from housing. The council decided to proceed with redevelopment proposals. This reflected, in my view, the careful and considered reaction of the experts and the councillors involved in making decisions about new developments. The only cloud on the horizon of this example came to my attention in October this year. I learnt that elsewhere in the north-west the HSE had advised refusal of three applications for small-scale housing infill near to a hazardous installation. I learnt that the distances involved were about the same as my Halton example and the type of potential major hazard was similar. It seems to show, at first sight, an inconsistency or perhaps the application of different standards.

Bulk liquid sulphur dioxide storage
My final example relates to two planning applications, in each case for a new major hazard. In both cases the applications involved the bulk storage of liquid sulphur dioxide in 100- and 200-ton tanks. One was dealt with quickly and simply, the other became quite complicated. The first was simple, and was approved quite quickly, partly because the plant was well located topographically in relation to the likely dispersion routes of a heavier-than-air toxic gas cloud. The first case was also simplified because the applicants had completed most of the stages involved in carrying out a hazard and operability study. The second case on the other hand was difficult because the topography was flat and no study had been carried out. Permission was granted subject to a reduction in the maximum size of each liquid sulphur dioxide tank to 30 tons and also a reduction in the total amount held on site as well. The reason for this was partly because no detailed hazard review had been done

by the company and partly to reduce the maximum credible incident to a point where possible loss of life in nearby houses about half a kilometre away was unlikely. The council's consultants produced a hazard analysis of the likely dispersion of the toxic gas cloud in the event of a tank rupture and the council used this as a basis for coming to its decision. The HSE had no comments to make on this proposal before the council had to make its decision.

Conclusions

I have tried to paint a quick picture to you of the Halton area and its council and the atmosphere in which decisions are taken. Whilst the area is rather special, the decision-making processes and responsibilities are broadly the same as any other local council.

I have also given you a number of specific examples that contain a number of interesting lessons. The one lesson I wish to highlight is that my council has a practical problem with the Health and Safety Executive when dealing with major hazard proposals. My council lacks confidence in the advice received. That was one minor factor in 1974 in employing specialist consultants. It has been a major factor in their continued re-employment. Central government advice is not adequate for my council's purposes.

The acceptability of a particular risk to life is only one of many planning issues involved in deciding whether or not to permit a new proposal. It is a complicated balance of factors in which hazard and risk assessment must be weighed along with many other matters.

When it is realised that the assessment of hazard and risk in new developments is just one part of a wider process where the politicians are responsible for saying 'yes' or 'no', then my role, that of the HSE and that of any other expert is clear. We are advisors. If planners, safety assessors and other experts are advisors to the councillors and the councillors think that the advice is inadequate, who is wrong, and who should change?

In my introduction I said I wanted to identify to you the main problem in dealing with hazardous planning applications. I think the main problem can be identified, but the solution is not simple. My message is: we need a change of attitude by the HSE. I detect some change already but the organisation has a long way to go. It must be prepared to talk informally, early on, to industrialists and councils about new developments. When a formal planning application is submitted HSE must be able to respond quickly. Planning applications must normally be decided in eight weeks. When HSE responds its advice must be full, open and analytical, not short and bland. It must be prepared to explain to officers, councillors and council committees the logic of its advice. It must expect occasional public furore about proposals. Local councils need and deserve open, clear and consistent advice, otherwise they finish up taking ignorant and unnecessarily irrational decisions.

Some people might say that more open public scrutiny of major hazards would bring such new developments grinding to a halt. That is always a possibility. When society is faced with the alternative of a remote risk from a potential major hazard balanced against the loss of so many of the consumer products which result from that hazardous process then I think society's decision-makers will usually be sensible. Some local councils may sometimes say no, even when faced with the full and open facts on a new proposal. The planning system allows for a right of appeal. In Halton, so far, we've never had that happen. Though it could happen in the future I think the record so far shows that local councils can cope in a considered and responsible manner when they are given full and open advice. Halton (and Cheshire) followed a special route by employing consultants, whereas the central government body that claims advisory responsibility in this matter should in my view provide adequate and open advice to all local planning authorities.

I feel I have been blunt. Perhaps I have been over-critical. Bernard Payne and I have talked about two aspects of the same problem. He concentrated on the need for more openness in the area of plan preparation. I have concentrated on the need for more openness when dealing with hazard and risk in planning applications. It just happens that the HSE is in the firing line but I would like to repeat what Payne said at the beginning of our chapter. Our comments are made constructively; we feel that open debate is a good way of improving the service that Parliament and local councils rightly demand of its expert advisors.

Appendix: suggested legislative changes

I consider there should be four changes:

(a) A clear and mandatory set of regulations defining major hazards (already in draft).
(b) Stemming from these regulations we need a clear and open list of specific hazardous installations.
(c) Having defined 'hazard' and having defined the sites, then planning authorities should be required to consult HSE within and around the adjoining hazardous installations.
(d) Having defined 'hazard' then such a use should be a special use for planning purposes, perhaps with no right to change the use within the use-class as well as in or out of it.

All four suggestions relate to new or amended statutory instruments, not new Acts of Parliament.

As far as defining the location of sites is concerned, they are all known now — only their boundaries are in doubt. As regards the seven sites in my district, I see no problem in definition.

The requirement to consult HSE is a simple matter of

insertion in a revised General Development Order, one of which is likely in 1980. HSE should not be given longer than one month to respond. The definition of types of planning application for consultation purposes will have to relate to adjoining land applications as well and the difficult problem of 2-kilometre zones etc. Again, I see a soluble problem, probably through national guidelines and local agreement between HSE and local planning authorities. It is a question of assessing the size of the hazard and its possible impact, the nature of existing and proposed urban development and the experience and confidence of local industry and the public.

The fourth point relates to altering the Town and County Planning Use-Classes Order. This statutory instrument defines, amongst other things, various categories of industrial use. A special use-class based on the HSE's definition of hazard can be introduced, just as the Alkali Acts provide the definition for an existing special industrial class. This would then establish existing 'hazard' sites for planning purposes. Changes in existing industrial areas to 'hazardous' uses would then need planning permission. Changes of 'hazard' within existing defined sites would either require planning permission or would more likely require sanction by HSE. It is usually the existence or non-existence of a hazard that is a planning issue. Thereafter, only a major change of scale of hazard would arouse new planning issues. In my experience this would usually need planning permission.

The recently published second report of the Advisory Committee on Major Hazards contains views and recommendations on planning (and other) matters. Like those in the first report, most of the points made are sensible and practical. The suggestion in para. 94 is particularly sensible, but para. 93's rejection of the Use-Classes Order seems to be based on a partial misunderstanding of its use. The introduction of a special class would be of great value, with or without an amendment to the definition of 'development'.

The second report mentions that it might be difficult to amend the definition of 'development' (para. 92). It is interesting to note that the Local Government Planning and Land Bill (No. 1) contained an amendment to this definition; this shows that an amendment can be introduced relatively easily.

6 *Risk and legal liability*

J. McLoughlin

Scope of this paper

This symposium concerns the acceptability of risk. In discussing risk and legal liability I have excluded any consideration of criminal law. By designating an act or consequence as criminal, society has indicated what it regards as unacceptable, but those laws are the fruits of arbitrary or morally or politically inspired decisions which would do little to answer the questions raised here. This paper will therefore deal only with civil liability.

Kinds of risk which are relevant to civil liability

For the purpose of considering risks in the context of potential civil liability at English law, it is convenient to classify them as follows:

(*a*) Risk that a certain kind of event, which is foreseeable, might occur, e.g. an escape of toxic gas to the atmosphere.

(*b*) Risk that a certain kind of damage might be done or injury be suffered e.g. pneumoconiosis, dermatitis.

The risk of damage might arise from an intentional act or practice, such as contact with a particular industrial lubricant, or from a foreseeable contingency arising under (*a*) above.

In deciding whether or not there is liability, the court takes into account not only the degree of risk that damage will be done, but also in conjunction with that the extent of the potential damage. In cases of personal injury that would be in terms of the number of persons put at risk and the seriousness of the injuries they might suffer.

(*c*)   Risk of consequences which could not have been foreseen, e.g. unforeseen effects of the use of a new pesticide.

Not all the above risks have a bearing on a court's decision in any one case, even where the risks undoubtedly exist. Different kinds of risk are taken into account according to whether liability is based on fault, or whether it is strict in the sense that simply because the damage is the consequence of a man's act he is to be held responsible. Fault liability and strict liability are therefore dealt with separately below.

Risk and fault liability

Intention, recklessness and negligence

Where liability depends on fault, that fault may lie in intention, recklessness or negligence. In this context the intention referred to is the intention that the damage which was done should in fact be done, and such intentional acts are therefore outside the scope of our present consideration. 'Recklessness' in this context means awareness that a particular kind of damage might be done, coupled with a blameworthy disregard of that possibility. 'Negligence' means a blameworthy inadvertence to such a possibility. In these senses, both recklessness and negligence may be involved in the creation of risk.

The difference between negligence and recklessness is therefore a foresight of possible consequences. Because of that foresight, the latter may be more blameworthy; but civil liability leads to compensation, and the measure of compensation depends on the extent and gravity of the damage or injury suffered, not on the degree of blame.[1] For our purposes, therefore, it will be sufficient to examine the elements of negligence, remembering that all save that of foresight apply to recklessness.

Foresight

A man cannot be made liable for negligence unless it can be shown that he should have foreseen that damage of a kind which was done might be done to the plaintiff as a consequence of his act.[2] In general, the foresight expected is that of the reasonable man, but anyone professing a particular competence is expected to have the foresight normally shown by a man of his trade, profession or recognised competence. Therefore a company manufacturing chemicals would be expected to have the foresight which would be given by the combined knowledge of the chemists, engineers and other experts normally engaged to establish and operate their processes, together with any other knowledge, such as that of the behaviour and effects of any gases they should foresee might be released to the atmosphere which is common to competent manufacturers of chemicals of that kind.

It will be noted that the damage of which complaint is made must be of a kind which was foreseeable. Therefore if an operator could foresee the possibility of an escape of gas, with obstruction of vision on the highway and resulting collision, he will not necessarily be liable for ill-health of a passer-by due to toxic effects which could not have been foreseen.

The problem of reasonable foresight is separate from the question as to what degree of care might reasonably be expected of a person. If I could not foresee an event, in other words if I was unaware of the particular risk, I could not be expected to take precautions. On the other hand I might foresee the possibility, but consider the risk so small that precautions could not reasonably be expected of me. For our purposes, however, they may be considered together, and indeed they often are in legal proceedings. There is no point in being aware of precautions that are not needed: conversely if a person professes to be competent to carry out a particular operation, he is usually, although admittedly not always, expected to be aware of those consequences which would be so serious that a reasonable man in his position would safeguard against them. For convenience the two problems are dealt with together. Therefore, the comments below may be taken to apply to the degree of foresight required to avoid liability as they apply to the degree of care expected.

The degree of care required by law
The degree of care required and the extent of the precautions to be taken in order to avoid liability, depend on a number of factors:

(a)   The degree of risk of a potentially damaging event occurring, and the risk of damage being done.
(b)   The extent or gravity of the damage which may be done.
(c)   The importance of the object to be attained by the risk-bearing activity.
(d)   The expense and difficulty of precautions which may be taken to reduce risk.

In an ideal world, the courts would reach decisions with both precision and overall consistency. Even within the vague limits of those terms, however, such an ideal is unattainable. It would involve a quantification of the risks referred to in (a), and an evaluation of those matters referred to in (b), (c) and (d). I do not know whether the first can be done: I do know that the second cannot, for so many subjective assessments are involved. Yet it is an objective towards which all concerned with justice will agree we should strive. For that reason the 'courts welcome any scientific quantification, and any economic evaluation; but by the same reasoning, in the end their decisions are based not on mathematics, but on a judicial sense of what is just in the circumstances. To a scientist that may sound too vague and subjective to be satisfactory, but clearly it is less dangerous than reaching by scientific means a firm decision on data which is inadequate in the sense that it does not take into account all the complexities of human activities, human relationships, and the variety of human values.

The matters referred to in (*a*) to (*d*) are dealt with separately below.

Because so many other factors are involved, it is not possible to spell out from the cases a degree of risk which the court would find acceptable.

At one end of the range of cases is *Fardon* v. *Harcourt-Rivington*,[3] in which a dog locked in a car jumped up against a rear window and broke it. A passer-by lost an eye as a result of being hit by a splinter of the glass. The owner was found not liable, the judge saying:

people must guard against reasonable probabilities, but they are not bound to guard against fantastic possibilities.

and

the user of the road is not bound to guard against every conceivable eventuality, but only against such eventualities as a reasonable man ought to foresee as being within the ordinary range of human experience.

At the other end is *Bradford* v. *Robinson Rentals*[4] in which a driver was sent on a journey of 450—500 miles in an unheated van with a defective radiator in the severe winter of 1963. The employer was liable for the frostbite suffered by the driver.

Of the cases between, it would probably be of interest to examine two cases in which figures were quoted. The first was *Haley* v. *London Electricity Board*.[5] The Board had excavated the pavement, and as a precaution against pedestrians falling into the hole had placed a punner hammer across the approach to one side of it.

The end of the handle rested on a railing two feet high, the head on the pavement. A blind man approached, his white stick passed over the hammer, and as he stepped forward the handle of the hammer caught his leg 4½ inches above the ankle. He fell into the hole and was injured. At the trial evidence was given that there were 7321 registered blind persons in the London area and 107 000 in Great Britain as a whole. One person in 500 was blind, and a great proportion of those often went out alone. The court found the Board liable, holding that it should have taken into account the possibility of a blind man passing that way.

The case raises the question whether the total number of people likely to pass that way is a relevant factor.

The other case was *Tremain* v. *Pike*.[6] In that case a farm worker contracted Weil's disease (Leptospirosis) in the course of his work, probably as a result of contact with rats' urine. There had been a considerable growth in the rat population on the farm that year, but the court was not satisfied that the employer knew or ought to have known that. Leptospirosis was present in 40%–50% of rats in this country, and in damp conditions remained active in urine for two or three days. Susceptibility of humans to the disease is regarded as very slight. From 1933 to 1948 948 people had contracted the disease (61.4 per annum) of whom 45 had been farm workers (2.8 per annum). The disease could be very serious, and in some cases had proved fatal. Against this background of fact the court concluded that the risk to the farm hand was 'very low indeed', and that his employer 'could not reasonably have foreseeen' that the man would catch the disease. Moreover, the provision of protective clothing would have been 'out of all proportion in cost and effort to the risk which had to be countered'. The employer was therefore found to be not liable.

A further principle of law is relevant here – that you must take your victim as you find him. In other words, the fact that the plaintiff was particularly susceptible to that kind of injury affords no defence. A workman was splashed on the lip with molten metal, and developed cancer there. It was found that he had had a 'pre-malignant condition' which made it more likely that the splash would cause cancer in him than in the normal man. This proved no defence for the employer.[7] As in Haley's case, therefore, any quantification of risk must take into account the number of people with relevant susceptibilities, abnormal though those people may be.

That the court takes into account the extent or the gravity of the damage which might be done is well illustrated by *Paris* v. *Stepney Borough Council*.[8] A workman who, to the knowledge of his employer, had only one good eye, was removing a U-bolt from the back axle of a vehicle. He was not wearing goggles, and when he struck the bolt a chip of metal flew into that good eye. In holding the employer liable the court made it clear that there was 'no reason for excluding as irrelevant the gravity of damage that the employee will suffer if an accident occurs.'

On the other hand, the greater the importance of the venture in which a person is engaged, the greater will be the latitude he is allowed in creating risk. This was taken into account when a fireman was subjected to risk and suffered injury when going to assist a woman trapped under a heavy vehicle. 'You must balance the risk against the end to be achieved.'[9]

Finally, the court takes into account the expense and difficulty of the precautions which would have to be taken to reduce the risk. Where the only precaution against workmen slipping on a wet oily floor would have been the temporary closure of a substantial part of a factory employing 4000 men, the employer was held entitled to subject the men to that risk.[10]

Causal relationship

There must, of course, be a causal relationship between the act of the defendant and the damage suffered by the plaintiff. The standard of proof, however, is the balance of probabilities, and circumstantial evidence may be sufficient.

A workman was employed cleaning out brick kilns, and was exposed to clouds of abrasive dust. No adequate washing facilities were provided and he had to cycle home caked in sweat and grime, and still sweating from his continued exertion. When the workman contracted dermititis the employer was found liable. Exactly how the disease occurs was not known to medical science, but it had been established that the risk of it occurring was substantially increased by the circumstances. The dictum of Lord Wilberforce in this case is particularly relevant to the subject matter of our discussions 'Where a person, by breach of duty or care, created a risk, and injury occurs within that area of risk, the loss should be borne by him unless he shows that it had some other cause.'[11]

Damage which is too remote a consequence

The line of responsibility is not endless; on grounds of policy a cut-off point is established. Thus, even within the bounds of foreseeability, there are some consequences which are considered too remote to require compensation. When a workman is injured and laid off work, the Treasury loses revenue, but no-one suggests that it has a cause of action for the revenue it has lost.

Risk and strict liability

Strict liability in English Law

A person who bears strict liability for an act is liable for consequent damage even though he may not have been at fault. This has two consequences for him of major significance if he is engaged in activities which generate risk:

(a)   The plaintiff does not have to prove fault in order to
      succeed. In view of the cost and uncertainty of litigation,
      this means that in the event of damage being done, legal
      actions are far more likely to be brought. Moreover, in
      many cases evidence of how and why the damaging event
      occurred is not available, perhaps because all material
      evidence has been destroyed in the event itself. In the
      absence of such evidence the plaintiff who has to prove fault
      loses:[12] where liability is strict, in the absence of such
      evidence he succeeds.
(b)   The elements of foresight and duty of care necessary to
      fault liability do not necessarily apply where liability is
      strict. To that extent the defendant takes the consequences
      of the risks he creates.

It must be added that 'strict liability' is a general term
covering varying degrees of strictness. The range of
events or damage for which a man is strictly liable may
be more or less limited; and there may be a wide or
narrow range of defences available to him. Such defences
may protect him from the consequences of the deliberate
or negligent acts of the plaintiff himself, acts of
terrorism, or natural phenomena which could not have
been foreseen or against which adequate safeguards
could not have been taken.

At common law the rule in *Rylands* v. *Fletcher* im-
poses a form of strict liability on persons who keep
hazardous substances on their premises.

the person who for his own purposes brings onto his lands and
collects and keeps there anything likely to do mischief if it
escapes, must keep it in at his peril, and, if he does not do so, is
prima facie answerable for all the damage which is the natural
consequence of its escape.[13]

The rule does not apply to natural uses of the land[14]
such as agricultural uses, and even the extraction of
minerals,[15] but in all normal circumstances it applies
to industrial activities including the storage of oil,
hazardous chemicals and gases.[16]

Several statutes impose strict liability. Under the
Animals Act 1971, the keeper of a dangerous animal is

strictly liable for damage that it does; there is the very strict liability of operators of nuclear installations under the Nuclear Installations Act 1965, and strict liability for owners and masters of marine oil tankers under the Merchant Shipping (Oil Pollution) Act 1971.

Present trends

In general, the English courts have shown themselves unwilling to extend the application of strict liability, despite the fact that the principles underlying the rule in *Rylands* v. *Fletcher* appear to be apposite elsewhere.

Parliament, on the other hand, in recent years has imposed strict liability in some areas of known risk, apparently due more to the effect of international agreements than to any other single cause. Of the statutes mentioned above, the Animals Act 1971 placed on a statutory basis a strict liability already imposed by common law, but the other two Acts introduced strictness where it did not exist before, to bring into effect the terms of international conventions.

Probably the most marked, and to us very relevant, extension of strict liability is that for the effects of harmful or defective products. Recent years have seen a hardening of the attitude of the courts in this respect in the United States, France, West Germany, the Netherlands, Sweden and Switzerland. In England the common law remains unchanged, but the Consumer Safety Act 1978 has introduced strict liability for breach of safety regulations concerning consumer products. There is little doubt that there will be further developments in this field.

One, which in one form or another we can confidently expect, will affect most of our manufacturing industry. There is already a draft EEC directive which seeks to impose strict liability for the effects of defective or harmful products for a period of ten years from the date of distribution, even though at that date the producer could not have known of the defect or harmful characteristics, and whether or not the article could have been regarded as defective or harmful in the light of scientific and technical developments at that time.

The relevance of strict liability to risk creation

The principle of Rylands v. Fletcher

It has long been considered by the courts that the incidents of everyday life create minor risks which the ordinary man is prepared to accept. As one judge stated,[17]

In the crowded conditions of modern life, even the most careful man cannot avoid creating some risks and accepting others. What a man must not do, and what I think a careful man tries not to do, is to create a risk which is substantial.

In addition there may be exceptional risks which emerge through the fault of no man — the loss must then be borne by the unfortunate person on whom it falls. That is the basis of the general rule of no liability without fault.

On the other hand, the *Rylands* v. *Fletcher* man[18] creates exceptional risks. Not only has he brought onto his land something which might be dangerous if it escapes, but he has departed from the natural use of land. The justification for imposing strict liability on him was best expressed by one judge when he said[19] 'It must be some special use bringing with it increased danger to others, and must not merely be the ordinary use of land'. Admittedly he went on to say 'or such use as is proper for the general benefit of the community' but that appears to introduce another and less acceptable principle. Perhaps that is merely a characteristic of a war-time decision.[20]

It is submitted that the introduction of exceptional risks, against which the ordinary precautions which the individual can take are insufficient safeguards, justifies the imposition of strict liability. This surely lay at the basis of the conventions which demanded strict liability for the operators of nuclear installations and the owners of tankers which carry oil in bulk. It is submitted also that the application of the principle could be extended to cover other risk creating ventures, such as the distribution of new pesticides and other potentially virulent products.

Strict liability and known risks

Even when the danger is not exceptional, however, there is still a case for imposing strict liability. This case becomes clear whenever the risk is quantifiable. It is best presented by means of examples.

A representative of a major oil company has claimed that there is a major pollution accident from marine oil tanker traffic once every three thousand tanker years.[21] For every tanker at sea for three thousand years there is on average a major spill. With three thousand tankers at sea, that gives one major spill every year. Thus the oil companies create a quantifiable risk and an anticipated consequence. That risk is created by a venture entered into for their own purposes. Surely a pollution accident is thus a foreseeable consequence just as much as the results which flow from the act of a careless man. Borrowing from the law of negligence,[22] are we not justified in saying that the companies should bear responsibility for the foreseeable damage done as a result of actions voluntarily performed? Parliament apparently thought so when it passed the Merchant Shipping (Oil Pollution) Act in 1971.

The same reasoning can be, and has been, applied to blow-outs at offshore drilling rigs. It has been observed that there is one blow-out for about every 500 drillings. We are now due for another in the North Sea. It may well be an 'accident' in the sense that there was no lack of due care on the part of the operator. Yet by his operations he created the risk, and the occurrence of another blow-out was foreseeable. Strict liability seems justified, and indeed the terms of the London Convention of 1976 require it.

All this points, however, not to strict liability for the individual, but to collective liability on a no-fault basis. In so far as those who are involved insure, collective liability is achieved: and subject to exceptions and special defences, the no-fault basis now seems to be gaining more general acceptance.

The same reasoning can again be applied to the less exceptional risks of road accidents. When those at fault have borne their responsibilities, there remains a residue of deaths and injuries which have not been attributed to the fault of any known person, and some which cannot be so attributed. The problem is complicated by the varying degrees of skill and care exercised by motorists, but the principle outlined above seems to have some application here. This, combined with other reasons not relevant to our discussion here, suggests that, subject to prior responsibility for any fault, strict liability might reasonably be borne by motorists collectively.

It is relevant to note also that although under strict liability the individual or his insurer bears the initial responsibility of compensating for the damage done, where that damage is done in the course of a trade or business the cost of providing for compensation is passed on in the price of goods or services. This means that ultimately those who enjoy the benefits of the enterprise pay for the damage it causes. That is surely right in principle.

Liability for acts of strangers
In many forms of strict liability there are exceptions for acts of strangers, at least where such acts could not have been anticipated by the defendant. This is true of the rule in *Rylands* v. *Fletcher*,[23] and of the Merchant Shipping (Oil Pollution) Act 1971 where the stranger intended to do damage.[24]

The Nuclear Installations Act 1965 makes the operator of an installation strictly liable for the consequences of a nuclear incident except, *inter alia,* where the incident occurred during the course of armed conflict.[25] This absolves the operator from the duty to compensate for damage caused by terrorists, and this appears to apply even if he has taken no precautions against them.

It is not exceptional today to make a person liable for the acts of even the negligence of others.[26] For some enterprises, intervention by terrorists is a known risk. It is unlikely that the terrorists themselves can be made to pay compensation, and the question arises — who is to bear the loss? In some circumstances the police authority may be made liable under the Riot (Damages) Act 1886, but there are many cases in which the Act will not apply.

Oil companies, in particular, are concerned for their oil rigs. They are willing to accept strict liability for the consequences of offshore operations in general, but strongly resist inclusion of the consequences of terrorist action. The result of that, of course, is that other innocent people will bear the loss.

There is much to be said for making the operator of a dangerous installation which may attract terrorist attack strictly liable for such consequences. In the first place he presents the target and thereby creates that particular risk. Secondly, subject to protection provided by public authorities, he is the only one who can take such precautions as are possible and thereby reduce the risk. Thirdly he is the best placed to take out insurance cover, in so far as insurance is available. Finally, he is the only medium through which the burden can be passed to those who enjoy the benefits of his operations.[27] Alternatively the State, of course, as a matter of policy could provide both protection and some insurance cover,[28] and there is a very strong case for the community as a whole sharing the loss.

Unknown risks

So far the discussion has been about types of risks
which are known, and in many cases quantifiable. There
are operations and products, however, which we cannot
confidently say are safe, and whose use we must admit
may create dangers as yet unidentified.

New pesticides provide an example. Some are potentially very dangerous, being based on the same substances
as those used in nerve gases prepared for chemical war-
fare. The producers of those pesticides subject them to
rigorous investigation supervised by the Ministry of
Agriculture before putting them on the market. Nevertheless, the producers themselves admit that, rigorous
though the checks may be, they still leave an area of
doubt. In these circumstances the producing firm, for its
own commerical benefit, puts the substance on the
market. They thus create an element of risk, and one
against which the general public cannot take any
adequate precautions. On the basis of the reasoning used
in the preceeding section of this paper, it is submitted
that strict liability for the consequences of risks arising
from that area of doubt would be justified. Which
appears, in this case at least, to justify the term of the
EEC draft directive noted above.

Notes

(1) Apart from the possibility of an award of punitive damages.
(2) *Overseas Tankship (U.K.)* v. *Mort's Dock Engineering (The
    Wagon Mound)* [1961] A.C. 388. *Hughes* v. *Lord Advocate*
    [1963] A.C. 837.
(3) (1932) 146 L.J. 391.
(4) [1967] 1 All E.R. 267.
(5) [1964] 3 All E.R. 185.
(6) [1969] 3 All E.R. 1303.
(7) *Smith* v. *Leech Brain* [1962] 1 Q.B. 405.
(8) [1951] A.C. 367.

(9) *Watt* v. *Hertfordshire County Council* [1954] 1 W.L.R. 835. The quotation is from the judgement of Denning L.J.

(10) *Latimer* v. *A.E.C.* [1952] 2 Q.B. 701. See also the comment in *Tremain* v. *Pike* quoted above.

(11) *McGhee* v. *N.C.B.* [1972] 3 All E.R. 1008 at 1012. See also *Gardener* v. *Motherwell Machinery & Scrap* [1961] 1 W.L.R. 1424.

(12) This is subject to the application of the doctrine of *res ipsa loquitur,* (the thing speaks for itself) where the circumstances raise a presumption of negligence.

(13) *Rylands* v. *Fletcher* (1866) L.R. 1 Ex 265 at 279–80. Affirmed by the House of Lords (1868) L.R. 3 H.L. 330.

(14) This limitation is discussed on pp. 116–17.

(15) *Rouse* v. *Gravelworks Ltd.* [1940] 1 K.B. 489.

(16) The case of *Rylands* v. *Fletcher* itself concerned the escape of water from a mill lodge.

(17) *Bolton* v. *Stone* [1951] A.C. 850 per Lord Reid at 867.

(18) See p. 114.

(19) *Rickards* v. *Lothian* [1913] A.C. 263 at 280.

(20) There is another judgement in which it was doubted whether the running of a munitions factory in time of war was a non-natural use of land. See *Read* v. *Lyons* [1947] A.C. 156 at 169–170.

(21) This figure was given in 1972. It was probably not correct then, and is most probably not correct now. Yet it will serve for the example.

(22) See p. 115.

(23) See *Rickards* v. *Lothian* [1913] A.C. 263 and *Hale* v. *Jennings* [1938] 1 All E.R. 579.

(24) See (b), pp. 106–7.

(25) See S. 13(4)(a).

(26) See *Hale* v. *Jennings* above. As to negligence see *Levene* v. *Morris* [1970] 1 All E.R. 144.

(27) See p. 118.

(28) For insurance cover provided by the State, see the Nuclear Installations Act 1965 S.18.

7 Risk, value conflict and political legitimacy

Stephen Cotgrove

In his Dimbleby Lecture on the risks of nuclear power, Lord Rothschild referred to environmentalists as eco-maniacs and econuts. In an address reported in *The Times* (12 October 1978), Paul Johnson referred to the ecological lobby as 'simply irrational'. On the other side of the debate, the Spring (1978) issue of *The Ecologist*, in an article entitled 'Reprocessing the truth', concluded that 'reason and truth no longer prevail at Public Inquiries'. Not surprisingly, the statement concluded that the only course open is a programme of non-violent disobedience.

The examples are familiar and can be multiplied. The debate on a whole spectrum of environmental issues abounds with charges of irrationality, sentiment and emotion — on both sides. Too often the protagonists face each other in a spirit of exasperation, talking past each other with mutual incomprehension. It is the dialogue of the blind talking to the deaf. This then is the problem. How can we account for the fact that intelligent and well-informed people face what appear to be major problems of communication, understanding and rational debate about environmental issues in general, and risk in particular? And what are the political consequences of such a crisis of rationality?

Fact, value and rationality

The crux of my argument is that problems of pollution and danger are always problems of value as well as fact. And I want to go further and suggest that conflicts of value present especial difficulties for our predominantly scientific and technocratic modes of rationality. Let us consider for a moment the notion of pollution. Now this is not a scientific term at all. Pollution involves a judgement, a threat to some standard of purity. So whether traces of fluoride in drinking water are defined as a dangerous pollution or a health-giving medication involves an evaluation and a judgement. Pollution is a social and political concept, not a scientific one. So are the concepts of risk and safety. It is always a question of acceptable levels. And acceptability involves an evaluation or judgement which goes beyond the 'facts', however problematic these may be. Society (or some section of it) decides that the existence of a substance in an environment above a certain level constitutes a danger to society. It is this which is defined as pollution. As Mary Douglas (1970, 1972) has shown so persuasively, rules against pollution can only be understood as part of the defence of a specific social order. And this is at heart always a moral order — some state of society which is deemed to be valuable and worth preserving. So when protagonists on some environmentalist debate appear to be arguing about some objective condition of, for example, rivers and waterways, more often than not they are arguing in defence of different moral and social orders.

This can be illustrated from the results of a comparative study on beliefs about the environment which we are carrying out at Bath. We obtained the views of a sample of members of the Conservation Society and Friends of the Earth and compared these with replies from a sample drawn from the pages of *Who's Who of British Engineers* and the *Business Who's Who*, and with

Table 1. Perceptions of dangers to the
environment

Damage scale score	Mean
Environmentalists	80.58
Public	72.21
Industrialists	58.68

a random sample of the general public. The questions
were a series of Likert-type statements from which we
were able to construct a scale of ten items, such as
'Rivers and waterways are seriously threatened with
pollution'. Industrialists scored much lower than either
of the other groups (Table 1).

Now it could be that the groups differ in the amount
of information they have, that the environmentalists are
better informed or more aware. But that hardly explains
the relatively high score of the public compared with
the low score of the industrialists.

I would like to suggest that the explanation is closer
to that put forward by Mary Douglas. Quite simply, the
social and moral order favoured by the industrialists is
really very different indeed from that favoured by the
environmentalists. And this is why the industrialists
perceive less pollution and danger. The discharge of
effluent into rivers does not threaten their world of
wealth creation: it is seen as less of a danger to their
moral and social order. The level at which it becomes a
danger, and therefore defined as *pollution* is simply
different.

So, too, the debate between environmentalists and
their opponents over, for example, nuclear power, is
more than simply an argument about the effects of low-
level radiation, or the probabilities of accidents. It is
also a debate about the values which can justify nuclear
risks. Risk is not just a statistical calculation. It is also a
moral judgement about defensible conduct. For the proponents of nuclear power, the overriding importance of

wealth creation is sufficient moral justification for the risks involved. For the environmentalists, there is no such moral justification.

Competing paradigms

To make this line of argument convincing, it is necessary first to spell out just how different are the social and moral orders of the protagonists in debates over environmental issues. I want to argue that beliefs about environmental issues, about pollution, dangers and risks are embedded in complex systems of belief and value which constitute distinct cultures: that the way the individual sees the world, attaches meaning to his environment, evaluates risk, perceives pollution and polluters, is part of his culture. We can describe this as the anthropological problem. It is a commonplace that anthropological research faces the difficulty of understanding the meaning and significance of actions against the background of implicit cultural meanings which are not shared by the investigator. Beliefs are embodied in complex systems, which at the highest level of abstraction can be described as *world views* (weltanschauung). Even in everyday experience, it is sometimes difficult to understand another's point of view. When we ask, in puzzlement or amazement, 'why on earth did you vote for him?', we seek to elucidate the context of beliefs and values which will enable us at least to 'see', if not to agree with the reasonableness of the action from that point of view.

But the problems of communication and understanding in some circumstances can be formidable. Thomas Kuhn (1962) has used the concept of a paradigm to explore revolutionary changes in the ways in which scientists 'see' some part of nature. The switch from the phlogiston to the oxygen theory of combustion is one such example: or from Newtonian to Einsteinian physics. The Keynesian revolution in economics is another. It took a decade or two to convince treasury officials not to look at government expenditure in the same way that a prudent individual runs his personal economy: that in hard times, governments should spend and not save.

There are two points about paradigmatic views. Firstly, they generate major problems of communication. Indeed, Kuhn argued that opposing paradigms may be incommensurable, and defeat understanding. Secondly, to adopt a new paradigm is frequently akin to a conversion experience. Once you have come to see the world from the new viewpoint it is hard to switch and to imagine how you saw it before. It is analogous to the *gestalt* switch which occurs when a diagram of a staircase suddenly 'flips' from a top-down to a bottom-up perspective. Now, if the analogy can be applied to political beliefs, alternative social paradigms could draw attention to problems of political communication and understanding of such a magnitude, that they could threaten the legitimacy of the system. It is this possibility that we need to explore.

Until recently, there has been broad agreement on the overriding goal for society — to maximise economic growth, and the production of goods and services. The dominant issue at elections has been (and still is) the economic performance of the government and the promises of its challengers. There has, of course, been some disagreement about how economic goals can best be maximised, the fair and just distribution of the product, and the kind of power structure necessary to achieve this. But the dominant socialist critique of capitalism as formulated by social democratic parties has not constituted a radical challenge to the overriding goals and values of industrial capitalism (Parkin, 1968). There may be disagreements about the extent to which market mechanism may be subjected to social objectives and constraints. But in crucial areas, such as wage determination, powerful sectors within the labour movement have come out strongly in favour of free collective bargaining. 'Labourism', seeks to soften the sharp edges of the market place, and to cushion the weak against its harsher rigours. And it seeks to extend political control over the centres of economic power. It offers, in short, an alternative version of an essentially industrial society. Even the communist or state socialist versions do not differ in the centrality of economic values. They, too, bow to the imperatives of industrialisation, albeit in significantly different ways.

Now there are signs that this dominant social paradigm, which has gradually come to exercise a virtual hegemony since the industrial revolution, is losing its grip over the minds of many — and for a variety of reasons (Dahrendorf, 1979). We may be witnessing what Heilbronner (1976) has called the decline in the spirit of free enterprise. So, for example, Paul Johnson in a spirited defence of industrial capitalism to the Bank Credit Analyst Conference in the USA concluded that:

the free enterprise idea is losing, if it has not already lost, the intellectual and moral battle . . . the steady diffusion of ideas hostile to our free system continues remorselessly. Industrial capitalism, the free market system, is presented as destructive of human happiness, corrupt, immoral, wasteful, inefficient, and above all, doomed. Collectivism is presented as the only way out compatible with the dignity of the humane spirit and the future of our race. The expanded university system threatens to become not the powerhouse of western individualism and enterprise, but its graveyard. (*The Times*, 19 October 1978.)

What we are suggesting, then, is that the various specific values and beliefs on which environmentalists and industrialists disagree do not each exist isolated and unsupported, but are embedded in a more or less coherent world view. Moreover, this functions as an ideology, to legitimate and justify actions and policies. That is to say, we are suggesting that these constitute paradigms in the strong sense of the word, and that this may explain the exasperation with which those who hold alternative paradigms face each other in debate and the charges and counter-charges of irrationality and unreason. Industrialists and environmentalists, we suggest, inhabit different worlds. From where they each stand, the world looks different. What is rational and reasonable from one perspective is irrational from another. If the goal is maximising output, then nuclear risks are not only justified but it would be unreasonable not to take them. From another perspective, from the viewpoint of a quite different set of beliefs about how the world works, and quite different aims for some kind of more convivial society, to take even the possibly small, but in practice incalculable, risks for future generations stimulates a moral indignation which justifies unorthodox political action that crosses the threshold of legality. It is the taken-for-grantedness, the common-sensical character of opposing views which lies at the root of incomprehension.

It is obviously difficult to get across the compelling nature of alternative paradigms for their adherents. The flip from one paradigm to another can be akin to a conversion experience. Only those who have undergone such an experience can really appreciate its nature. The nearest many can come to such an experience is in literature, either read or enacted, in which the author succeeds in generating such a degree of empathy and identity with a character that we come to see and experience the world differently for a time. The differences between the competing paradigms are set out schematically in Table 2. But to capture their full flavour one needs to listen to the sometimes empassioned rhetoric of the true believers — on both sides.

The first, and it will be argued, most important difference centres around the core values. It is the value of wealth creation which is the master value of industrialism. In an attack on the environmentalists, C. C. Pocock, Managing Director of Shell International, made his priorities quite explicit:

Table 2. Competing social paradigms

	Dominant social paradigm	Alternative environmental paradigm
Core values	Material (economic growth) Natural environment valued as resource Domination over nature	Non-material (self-actualisation) Natural environment intrinsically valued Harmony with nature
Economy	Market forces Risk and reward Rewards for achievement Differentials Individual self-help	Public interest Safety Incomes related to need Egalitarian Collective/social provision
Polity	Authoritative structures: (experts influential) Hierarchical Law and order	Participative structures: (citizen/worker involvement) Non-hierarchical Liberation
Society	Centralised Large-scale Associational Ordered	Decentralised Small-scale Communal Flexible
Nature	Ample reserves Nature hostile/neutral Environment controllable	Earth's resources limited Nature benign Nature delicately balanced
Knowledge	Confidence in science and technology Rationality of means Separation of fact/value, thought/feeling	Limits to science Rationality of ends Integration of fact/value, thought/feeling

we see a huge agricultural dam . . . halted by environmentalists
when it was nearly finished, and then scrapped to protect some
obscure fish . . . It has been well said, 'The creation of wealth in
a world of want is a moral duty.' I suggest that that morality is
just as valid as the morality of the environmentalists.

The creation of wealth for the industrialists is a moral
imperative. And if wealth is the name of the game, then
the rules for winning that game follow: rewards for
enterprise, and risk, a free market, and creating climate
in which individuals are motivated to look after themselves and not to turn to others. Production and distribution require organisation and direction: the division of
labour is only possible within the framework of centralised hierarchically structured and imperatively co-
ordinated organisations. Respect for legitimate authority
— both at the industrial and political level — provides
the essential context of law and order within which the
forces of production and the free market operate. The
task of politics is to see that the game is played according to the rules but not to interfere with the players.

The alternative environmental paradigm differs on
almost every issue. The first and most obvious point of
difference is the environmentalists' opposition to the
dominant value attached to economic growth. This in
turn is reinforced by beliefs that the earth's resources
are finite — a view encapsulated in Boulding's telling
metaphor 'space-ship earth'. But their disagreement with
the central values and beliefs of the dominant social
paradigm runs deeper than this. Not only do they challenge the importance attached to material and economic
goals, they by contrast give much higher priority to the
realisation of non-material values — to social relationships and community, to the exercise of human skills
and capacities, and to increased participation in decisions
that affect our daily lives. They disagree too with the
beliefs of the dominant social paradigm about the way
society works. They have little confidence in science
and technology to come up with a technological fix to

solve the problems of material and energy shortages. And this is in part rooted in a different view of nature which stresses the delicate balance of ecological systems and possibly irreversible damage which may result from the interventions of high technology. They question whether the market is the best way to supply people with the things they want, and the importance of differentials as rewards for skill and achievement. They hold a completely different world view, with different beliefs about the way society works, and about what should be the values and goals guiding policy and the criteria for choice. It is, in short, a counter-paradigm, or adversary culture.

Acceptable risk

I would like to turn now to exploring the implications of this approach for policy and decision taking in rather more detail. To point out that policy on environmental issues must take account of values is in one sense now a part of the conventional wisdom (Ashby, 1978). It is recognised that there are competing priorities and that, for example, in choosing a site for the third London airport, or developing a policy for the control of pollution, building a new reservoir, or assessing the risks of nuclear power — values are at stake. The reservoir may involve a threat to plant life or amenity and this must be weighed against the loss of jobs or exports if water supply is not increased. Yet debates are frequently conducted as though what is at issue are questions of fact. Hence the exasperation of Lord Rothschild in the face of what he saw to be solid evidence that nuclear reactors are safer than windmills. This dominant mode of thinking tends to exclude values from its calculus simply because it assumes them to be non-problematic. We can describe this as the technocratic mode.

Technocratic mode

| Authoritative (scientific) facts, information | → | logical processing | → | rational decisions |

It operates within the dominant social paradigm, and takes for granted its core values and beliefs.

Now what is being argued here is that the technocratic mode is altogether too simplistic. It ignores what I have called the anthropological problem of meaning. It tends to focus on facts which can be measured and to exclude more intangible considerations. It assumes facts themselves to be less problematic than they are. And it fails to grasp the problems of understanding and communication which stem from holding different core or master values and their associated beliefs. The point can be illustrated by the researches of Otway and his associates (Otway, 1977; Otway, Maurer & Thomas, 1978) in Austria into public attitudes towards nuclear power. These findings are of considerable significance and are worth more detailed study. Firstly, it is necessary to make a distinction between a belief about an object, for example, that nuclear power will raise the standard of living and the evaluation of raising the standard of living as desirable. 'Attitudes' are the way in which beliefs and feelings combine. So a favourable attitude to nuclear power would require both the belief that it will raise the standard of living and a feeling that this is a good thing. Put very simply, beliefs X values = attitudes. Otway and his associates then distinguish between four main clusters of attributes:

(1) *Psychological risk factor* (e.g. exposing me to risk without my consent, and exposing me to risks which I cannot control).

(2) *Economic and technical benefits factor* (e.g. raising the standard of living and provide good economic value).

(3) *Sociopolitical risk factor* (e.g. leading to rigorous physical security measures and to dependency on small groups of highly trained specialised experts).

(4)   *Environmental and physical risk factor* (e.g. leading to water pollution or long-term modification of the climate).

The results from a sample of the Austrian public demonstrate the complex nature of the beliefs which contribute to public attitudes towards nuclear power. For those in favour, it was the technical benefits which contributed most to their attitude. For those against, it was risk aspects, both psychological and sociopolitical, which most accounted for their opposition. So, the tendency of the technocratic mode to concentrate on the economic benefits and environmental risks (which are the bases of *their* support) will fail to come to grips with the anxieties of the opponents which focus on sociopolitical and psychological risks.

What this study does is to underline the fact that the opposition of nuclear protestors goes beyond technical questions of risk and safety and economic benefits. It is the wider questions of social, political and psychological risk to which they attach importance. The significance and meaning of nuclear power for the social and moral order of the opponents is its promise of remote impersonal centralised bureaucracies, increased reliance on experts, loss of control over decisions which affect their lives, threats to personal liberty from the security requirements of the plutonium economy, as well as the risks of nuclear proliferation. To the supporters, it is the economic benefits which are of overriding importance. Their social and moral order is threatened by the failure to develop nuclear power. If the environment takes a knock or two or if society takes some calculated risks, then this is the price to be paid for the pursuit for the greater good.

Whenever we are involved in policy and decisions on risk and safety we need to move to a political mode of reasoning (see p. 134). This recognises that an adequate account of what it is to be rational and to act reasonably takes account of both beliefs and values. But even this model is dangerously simple. To operate effectively, it

Political mode

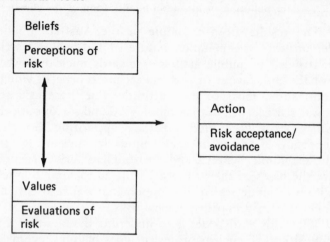

needs to be recognised that the perception of risk is itself an extremely complex process. Objective facts and evidence certainly come into it. But it is the meaning of the evidence which is important (Table 3). For example, in making a judgement on the reliability of the data, we will take account of its source. And here, environmentalists will frequently differ. They do not share the confidence of many in the advice of experts, and will be particularly cautious when those experts are employed by the protagonists in the debate. They are likely to

Table 3.

Perceptions of risk
How risky is it?
What dimensions/aspects of the object/action are relevant?
What is the source of data: how reliable/credible are the 'facts'?
What is the *meaning* and interpretation to be put on the evidence?

point out too that experts differ, and to be able to muster supporters with acceptable credentials whom they believe to be closer to the truth. Moreover, expert opinion itself undergoes change. Fresh evidence comes to light which undermines confidence in the dogmatic assertions.

Values for money: flowers or production?

When we come to deal with the value component in the equation, we move into even more troubled waters. Moral discourse presents particular difficulties both for the technocratic and for the related scientific modes. Judgements of morality are never simple matters of fact, however relevant facts may be. An 'ought' statement always goes beyond an 'is' statement. So this makes moral arguments especially difficult to prove and accounts for the exasperation which so often accompanies them. No cool detached analysis of observable facts or testing hypotheses will settle the argument for those who prefer the Beatles to Beethoven.

But there are deeper, more intractable problems of communication and understanding. Not only is it difficult to persuade people to change their values, but the anthropological problem of really seeing an issue in terms of a different set of values can be extremely difficult (Table 4). Take for example, the Cow Green reservoir, where the proposal to build a reservoir in

Table 4.

Evaluation of risk
Is it worth it?
What ends will it promote?
What are the relevant criteria/calculus?
What are the higher order/master values?

Upper Teesdale met with objections on the grounds that
it threatened the survival of what many botanists considered a unique community of rare plants (glacial
relict flora). The debate turned on the question of
weighing the benefits of preserving the flowers, against
the economic costs of the loss of employment and
export markets if the reservoir were not built. Lord
Leatherland put the issue nicely 'In my own simple
way I am asking whether I should decide between
flowers on the one hand, and people on the other —
people and their prosperity . . . I come down solidly
against flowers.' (Quoted in Gregory, 1971.) It may
look simple to the noble lord. But it is not. The crux
of the problem centres on the translation of one scale of
values to another: economic values against non-economic
values. There is no difficulty in translating degrees centigrade into degrees Fahrenheit. All you need to know is
the 'conversion' factor, or the relative value of one
degree versus another. The two scales are commensurable
and can be interchanged. Proponents of cost–benefit
analysis are trying to do the same thing. They say, if we
can discover what price people put on say the preserva-
tion of flowers, and then weigh that against the cost
they would sacrifice in the loss of production, then we
can come to a rational decision. But some scales of value
are not commensurable. You cannot translate degrees
centigrade into cubic centimetres. They are different
values: one is temperature and the other cubic capacity.
This can be put dramatically in calculations of the value
of human life. Economists have approached this in a
number of ways, for example, by capitalising the loss of
earnings. At first sight, the approach seems sensible
enough. But put the issue differently: ask a wife — how
much would you sell your husband for — or your child?
If every thing has a price, and can be reduced to an
economic calculus, then such a question would not
shock. It does. And you can't. That is what exasperates
the environmentalists: that the value which provides the

criterion is not one which they can accept. Cost—benefit analysis takes for granted that economic values and criteria provide the ultimate arbiter; that any action is justified if it pays, and is not justifiable if the economic costs are too high. Now not even the most enthusiastic defender of the market is prepared to go quite as far as that and to admit free trade in children or pornography. There are some circumstances in which higher order values override the imperatives of the market place, and into which cost is not allowed to enter. The environmentalists seeking to save the flowers are saying just this. But the supporters of the reservoir take it for granted as common sense that economic criteria are more important, and that if it is a choice between flowers or production, then the flowers must go.

The technocratic mode operates within the framework of agreed political values and seeks to depoliticise such issues by reducing them to an economic calculus. But amenity, beauty, or scientific truth cannot be quantified. Where some hold strongly to values which they see to be higher order values and these are not part of the taken-for-granted values of the technocratic consciousness, there is exasperation and emotion — on both sides — rooted in fundamental failures of understanding. The predominant scientific/technical rationality of modern industrial societies simply finds it difficult to come to grips with differences in values. There is really no vocabulary, no style of thinking within the dominant culture which provides a flower-value versus production-value calculus. On the contrary, the dominant technocratic mode dismisses such arguments as coloured by ill-informed sentiment, and therefore irrational. Even where it is recognised that there are conflicts of values there is difficulty in accepting these as a basis for rational decisions. Rationality is defined in such a way that it is reduced to an economic calculus. So, for example, in a scrupulously fair analysis of the Cow Green Reservoir debate, Roy Gregory (1971) concludes that those who took part in the debate reacted in predictable ways 'spontaneously and instinctively, and by reason of their background and interests, their hearts were on one side or the other... To judge from their speeches, several peers ... had adopted a more rational and open-minded attitude, asking themselves what really would be the cost to industry ... and ... the cost to scientific research' (p. 190). So, it is careful weighing of costs, not strong sentiments for the beauty of the countryside which for Roy Gregory characterises a rational decision. He too, cannot escape from the predominant economic calculus.

A crisis of legitimacy?

What are the implications of the problems for risk and environmental policy? How do the problems of communication and understanding affect the political and administrative processes of policy formation and decision-taking at all levels from local enquiries to parliamentary legislation?

The central point is that paradigms function as ideologies. They not only provide a map of where to go, and how to get there, but most important, they legitimate and justify courses of action. And it is here that disputes over environmental dangers pose one of their most serious threats to the political and administrative processes.

Legitimacy is more than legality. It is an acceptance of rules and decisions as in some way proper, which ought to be kept. There is a moral dimension to it. The bases for the conferment of legitimacy are complex. It rests in part in the belief that the procedures are fair and reasonable, that there is a genuine attempt to pursue the public good. It is when the channels of communication are clogged by incomprehension, or signs of deafness by the decision-takers and policy makers appear, that frustration and exasperation lends legitimacy and justification to unorthodox methods. And this is precisely what seems to happen in many disputes over pollution and danger.

Because of its taken-for-granted character, the dominant social paradigm can systematically repress the articulation of alternative viewpoints. Given support for economic values and growth, confidence in experts, and in the power of science and technology to come up with answers, then the conclusions of Mr Justice Parker at Windscale can be seen to be not only reasonable but right. Given the acceptance of the dominant goals and values of society, problems are seen to be essentially questions of means, soluble by harnessing knowledge and expertise to the political process. Rationality is defined in narrowly technical or instrumental terms. What are properly political questions involving conflicts of values and interests are depoliticised and treated as technical questions; what Habermas (1971) refers to as the 'scientisation of politics'. This, it is argued, is precisely what happened at Windscale. It is under such conditions that political institutions distort communications and there is no genuine dialogue. Hence the charges and counter-charges of unreason and irrationality between environmentalists and supporters of the status quo.

It is when political decisions are seen to be unreasonable, because they have failed to take account of alternative viewpoints, that there is a loss of confidence in legitimate processes, and a turning to illegitimate forms of direct action — disruption, boycotts, sit-ins and at the extreme, violence. And the danger then is that what is seen as a threat to law and order generates repressive responses which exacerbate the fundamental malaise.

Note

The researches reported here are supported by a grant from the Social Science Research Council, and assisted by Andrew Duff. Part of this chapter has appeared in Cotgrove & Duff, 1980.

References

Ashby, E. (1978). *Reconciling man with the environment.* Oxford University Press.
Cotgrove, S. & Duff, A. (1980). Environmentalism, middle class radicalism and politics. *Sociological Review*, 28, 2, May.
Council for Science and Society (1977). *The acceptability of risk.*
Dahrendorf, R. (1979). Towards the hegemony of post-modern values. *New Society*, 15 November 1979.
Douglas, M. (1972). Environments at risk. In *Ecology, the shaping enquiry*, ed. J. Benthall. Longman, Harlow.
Douglas, M. (1978). *Purity and danger.* Routledge & Kegan Paul, London.
Gregory, R. (1971). *The price of amenity.* Macmillan, London.
Habermas, J. (1971). *Toward a rational society.* Heinemann Educational, London.
Habermas, J. (1976). *Legitimation crisis.* Heinemann Educational, London.
Heilbroner, R. L. (1976). *Business civilization in decline.* Martin Boyars, London.
Kuhn, T. (1962). *The structure of scientific revolutions.* Chicago University Press.
Otway, H. J. (1977). Risk assessment and the social response to nuclear power. *Journal of the British Nuclear Energy Society*, 16 (4), pp. 327–33.
Otway, H. J., Maurer, D. & Thomas, K. (1978). Nuclear power: the question of public acceptance. *Futures*, 10 (2), pp. 109–18.
Parkin, F. (1968). *Middle class radicalism.* Manchester University Press.
Rothschild, Lord (1978). Risk. *The Listener*, 30 November 1978.

Index

Note

The index includes references to page numbers where a particular topic is to be found. However, the grouping does not include references to whole chapters dealing with a topic at length, e.g. planning. The contents list should be used in conjunction with the index as appropriate.